STOCHASTIC AND QUANTUM DYNAMICS OF BIOMOLECULAR SYSTEMS

STOCHASTIC AND QUANTUM DYNAMICS OF BIOMOLECULAR SYSTEMS

Proceedings of the
5th Jagna International Workshop

Jagna, Bohol, Philippines 3 – 5 January 2008

EDITORS
Christopher C. Bernido
M. Victoria Carpio-Bernido
Research Center for Theoretical Physics
Jagna, Bohol, Philippines

SPONSORING ORGANIZATIONS
Department of Science and Technology (Philippines)
Alexander von Humboldt Foundation (Germany)
Mindanao State University – Iligan Institute of Technology
Physical Society of Visayas and Mindanao

AMERICAN INSTITUTE OF PHYSICS

Melville, New York, 2008
AIP CONFERENCE PROCEEDINGS ■ 1021

Editors

Christopher C. Bernido
M. Victoria Carpio-Bernido

Research Center for Theoretical Physics
Central Visayan Institute Foundation
Jagna, Bohol 6308, Philippines

Email: cbernido@mozcom.com

L.C. Catalog Card No. 2008928313

ISBN 978-0-7354-0543-1
ISSN 0094-243X

Printed in the United States of America

CONTENTS

BIOLOGICAL STRUCTURES AND INTERACTIONS

INTERACTING SYSTEMS

Preface

The "*5th Jagna International Workshop: Stochastic and Quantum Dynamics of Biomolecular Systems*," organized by the Research Center for Theoretical Physics (RCTP), Central Visayan Institute Foundation, from 3 to 5 January 2008, served as a forum for interdisciplinary discussion on topics in the mathematical, physical and life sciences. Possible directions in the study of biomolecular systems were discussed in the light of recently developed analytical tools from physics and mathematics. The talks ranged from the pedagogical level (with graduate students and young Ph.D.'s as the target audience), review level, up to recent research results. The Workshop allowed ample time for informal, but intense exchange of ideas and perspectives.

The proceedings of the Workshop should reflect in part the flavor of the scientific meeting. The organizers are grateful to the speakers for submitting camera-ready articles for the proceedings. The volume starts with papers dealing with some mathematical methods in stochastic and quantum dynamics. These include infinite dimensional white noise analysis, the path integral method, stochastic processes with a repulsive drift of polynomial growth, and the WKB approximation. The second part of the volume presents applications to biophysical systems with articles on neurophysics. The papers are on nonlinear dynamics as a useful tool for quantitative analysis of electroencephalographic signals, the role of noise in neural processing, and a stochastic path integral approach to neurodynamics. The third part of this volume deals with biological structures and interactions. A pedagogical paper is followed by papers on the path integral modeling of protein conformations, the probability of polymer windings, and polymer loop conformations enclosing an area, and experiments on cell growth for mixed-cell culture. The last part of the volume deals with scaling laws and networks, and their application to modeling interacting social systems. The first paper of this part deals with violent conflicts where a sovereign nation can be compared to an organism, and an internal conflict compared to a disease. Since an organism fights a disease with its immune system, one expects similarities between social conflicts and immune reactions, including similarities in relevant scaling laws. This paper is then followed by one on bipartite networks as a useful tool for investigating interaction networks, and a paper using the random walk approach in investigating spatial networks of human settlements.

The Workshop was also a special occasion and opportunity to honor two distinguished scientists: Prof. Dr. Ludwig Streit and Prof. Dr. Frederik W. Wiegel, in appreciation of their scientific contributions to physics, mathematics and biophysics, many years of mentoring and scientific collaboration, as well as their unstinting support to the RCTP as members of its Scientific Advisory Board.

We first met Prof. Streit formally, during the second international conference on Path Integrals from meV to MeV, held in 1985 at the Center for Interdisciplinary Research (ZiF) of the University of Bielefeld, Bielefeld, Germany. Being then the

Director of the ZiF, Prof. Streit hosted the event. A former Editorial Board member of the Journal of Mathematical Physics, among others, Prof. Streit is a physicist with a passion for mathematics. Together with Professor Takeyuki Hida of Nagoya, Japan, he used insights from the Hida white noise calculus to formulate a rigorous interpretation of Feynman integrals. Prof. Hida had earlier introduced the white noise variables (velocity of Brownian motion) as coordinates of an infinite dimensional space. Later, Prof. Streit became our scientific host for Alexander von Humboldt research fellowships (1987-89 and 1996-97). The math-physics encounters in Bielefeld, and later in Madeira, nurtured mutual interest in deeper studies on the meaning and broader application of Feynman path integrals.

On the other hand, we met Prof. Wiegel, a plenary speaker during the third conference on Path Integrals from meV to MeV hosted in Bangkok, Thailand in January 1989 by Prof. Virulh Sa-yakanit. Even at that time, Prof. Wiegel was interested in biophysics having had noted works in chemoreception and the topology of polymers. He was then looking at applications of Feynman path integrals. Shared interests motivated more encounters, in Bangkok, in Bielefeld, and in Manila. He kindly extended support in 1992 when we started a small research center as a private initiative in an island province in the southern part of the Philippines aimed at fostering physics research in the country.

In 1995, during the inaugural Jagna Workshop, there was a further convergence of paths. Profs. Streit and Wiegel came to Jagna, Bohol for the first time. There have been subsequent visits since then. Indeed, they have attended all five of the triennial Jagna workshops. This is not easy considering the great gap between Eastern and Western cultures, felt most especially in the earlier years. This is not only in terms of the outlook and quality of life brought about by physical conveniences (or inconveniences), but also the longer time frames of project development in a developing country like the Philippines. In a happy fusion of physics and friendship, the seeds of interdisciplinary and cross-cultural endeavors scattered in Bielefeld and Bangkok have germinated in Jagna, in Manila and Iligan, just as they have in Madeira, Lisbon, Enschede, Amsterdam, Tokyo and Nagoya. It is, therefore, a great pleasure to honor Prof. Streit and Prof. Wiegel with other friends on the occasion of their 70th birth year.

The Editors

MAJOR SPONSORS

Department of Science and Technology (Philippines)

Alexander von Humboldt Foundation

Mindanao State University-Iligan Institute of Technology

Samahang Pisika ng Visayas at Mindanao

WORKSHOP ORGANIZERS

Christopher C. Bernido **M. Victoria Carpio-Bernido**

The Workshop Organizers are grateful to

Dr. Graciano P. Yumul, Atty. Jesus Antonio Z. Carpio, the Central Visayan Institute Foundation faculty and administrative staff, the Bernido and Carpio families,

for their invaluable support.

5th Jagna International Workshop: Stochastic and Quantum Dynamics of Biomolecular Systems

LEFT to RIGHT:

1st Row: G. Concepcion, E. Melchor, J.P. Esguerra, C.C. Bernido, A. Albano, M. Oliveira, M.V.C. Bernido, L. Streit, F.W. Wiegel, J.L. da Silva, J. Klauder, N. Cue, H. Ezawa, Ph. Blanchard, T. Hida, M. Grothaus, T. Nakamura, A.R. Nasir

2nd Row: J. Ando, J. Magallanes, J. Teo, E. Cuansing, Jr., R. Solidum, P. Macansantos, E. Jimenez, J. Bornales, C.J. Gravino, P.A. Ramirez, C. Patacsil, G. David, W. Alangui, B.V. Gemao, P. Galos, R. Reserva, D. Sauquillo, R. Rañoa

3rd Row: M. Evardo, M. Pagulong, J.P. Baja, G. Acas, A. Tadle, J. Cuñado, M. Cagas, N. Lindao, L.G. Villaruz, L. Bastatas, E. Gravador, J.V. Mendez, M.D. Dumaluan, R. Loberternos, N. Lasta, V. Galbizo

4th Row: J. Salas, R.M. Roxas, M.G. Bato, N. Acasio, H. Aringa, M. Confesor, K. Gargar, R. Vitancol, C. Monterola, J. Dela Cruz, R. Gania, K.J. Carpio, M. Nakamura, F.R. Bagsican, C. Soriano, A. Aceron, J. Bagaipo

MATHEMATICAL METHODS IN STOCHASTIC AND QUANTUM DYNAMICS

Infinite Dimensional Spaces in Stochastic Analysis

Takeyuki Hida

Nagoya, Japan

Abstract. A review of the notion of generalized white noise functionals is presented and the concept of infinite dimensional spaces of random functions is discussed.

Keywords: White noise functionals, Brownian motion, infinite dimensional space

PACS: 02.30.Sa, 02.50.Ey

INTRODUCTION

The space of generalized white noise functionals is the basic tool of white noise analysis, and the space plays dominant roles in the analysis. Unfortunately, we have often met serious misunderstanding of generalized white noise functionals, to our regret.

The aim of this note is two-fold. One is to recall the definition of the *generalized white noise* functionals beyond the intuitive interpretation. The other is to discuss the concept of *infinite dimensional spaces* of random functions.

As for the motivation of the first aim, we did realize that an intuitive interpretation of generalized white noise functionals, in particular white noise $\dot{B}(t)$ itself, was not so well interpreted. For example $:\dot{B}(t)^2:$ the renormalized square of white noise $\dot{B}(t)$ is often misunderstood as if it could be reduced to a functional of Brownian motion $B(t)$ in the classical sense. We have to give better explanation on how to define generalized white noise functionals in the next section. Actually, the notion did appear already in 1975 in the Carleton Math. Notes [1] and has often been referred to in quite many papers.

We are going to present an elementary method of understanding the notion of generalized white noise functionals. Heuristically, we intended to start with the step, the so-called *Reduction* for the investigation of random complex systems. A realization of this idea

CP1021, *Stochastic and Quantum Dynamics of Biomolecular Systems,*
Proceedings of the 5th Jagna International Workshop, edited by C. C. Bernido and M. V. Carpio-Bernido

is first to construct independent random variables out of the given random phenomenon, in such a way that the collection of those independent random variables should contain the same amount of information as the given system, where the causality in time is always taken into account. Such a system of independent random variables is hopefully atomic. Thus, the given system would be expressed as a (nonlinear, in general) function of the independent random variables which are just constructed.

We now wish to make a short note. It was 1994 when I met Professor John Klauder on the occasion of the international conference on "Group theoretic methods in Physics" held at Osaka. I explained the idea of Reduction to him. Good examples are the white noise, that is the time derivative $\dot{B}(t)$ of a Brownian motion $B(t)$ and Poisson noise $\dot{P}(t)$, where $P(t)$ is a Poisson process, and so forth. He understood my idea and kindly suggested to me to call such a system of independent random variables *system of idealized elementary random variables.* Since then, I have been very happy to use these terms; often modified to say *idealized elemental random variables.*

We then come to a review of idealized elementary random variables with somewhat better interpretation than before. Once a system of variables has been fixed, we have been naturally led to introduce generalized white noise functionals. There are two methods of introducing the space of generalized white noise functionals. One sticks to polynomials in $\dot{B}(t)$'s and Hilbert space structure is kept. The other is an infinite dimensional analogue of the Schwatz space of generalized functions on R^n. We are then ready to carry on the analysis of generalized white noise functionals.

The last section will be devoted to a discussion of the notion "Infinite Dimensional Space" as an application of the results in the last two sections. Such a consideration is interesting by itself and should be kept in mind when we discuss applications.

GENERALIZED WHITE NOISE FUNCTIONALS (A REVIEW)

The steps of how to understand white noise $\dot{B}(t)$ are as follows.

1). In classical stochastic analysis, $\dot{B}(t) = \frac{d}{dt}B(t)$, where $B(t)$ is a Brownian motion, has only formal significance. But, stochastic integral makes sense: it is of the form

$$\int f(t)dB(t) =< \dot{B}, f >= \dot{B}(f).$$

First it is well defined for f to be a test function. We understand that it is a continuous bilinear form. Then, it ends to be a continuous linear functional of f with $f \in L^2(R)$.

Thus white noise is accepted only after it is smeared.

2). I.M. Gel'fand gave a definition of generalized stochastic process in 1955. White noise is a generalized Gaussian process with characteristic functional

$$C(\xi) = \exp[-\frac{1}{2}\int |\xi(u)|^2 du],$$

where ξ is a test function. Thus a white noise, as a generalized stochastic process, is represented by $\dot{B}(\xi)$, namely we have a family of (ordinary) random variables parameterized by $\xi \in E$, where E is a nuclear space. It is continuous in ξ.

3). Generalized linear functionals.

In step 1) we have established an isomorphism

$$L^2(R^1) \;\cong\; H_1 = \{\dot{B}(f), f \in L^2(R^1)\}.$$

The ideal random variable $\dot{B}(t)$ corresponds to the case where the test function ξ is replaced by the delta function δ_t which is in the Sobolev space of order -1, denoted by $K^{(-1)}(R^1)$ which is the dual space of $K^{(1)}(R^1)$ of order 1. We have the inclusions

$$K^{(1)}(R^1) \subset L^2(R^1) \subset K^{(-1)}(R^1),$$

where the mapping from left to right is continuous injection.

First we restrict the isomorphism between $L^2(R^1)$ and H_1 to

$$K^{(1)}(R^1) \;\cong\; H_1^{(1)}.$$

Actually $H_1^{(1)}$ is defined so as to hold the above isomorphism. The dual space $H_1^{(-1)}$ of $H_1^{(1)}$ can be defined and we have

$$H_1^{(1)} \subset H_1 \subset H_1^{(-1)} \tag{1}$$

in pararell with

$$K^{(1)}(R^1) \subset L^2(R^1) \subset K^{(-1)}(R^1).$$

A member of $H_1^{(-1)}$ may be written as $\dot{B}(g), g \in H^{(-1)}(R^1)$ which is an extension of a stochastic bilinear form. Now we are happy to take g to be a delta function δ_t to have $\dot{B}(\delta_t) = \dot{B}(t)$.

From now on, we understand $\dot{B}(t)$ in the sense established above. Members in $H_1^{(-1)}$ are called linear *generalized white noise functionals*. We may say that *identity* of $\dot{B}(t)$ is given as a member of $H_1^{(-1)}$.

Having had the system of varibles $\{\dot{B}(t)\}$, which now have concrete meaning, we can establish a class of acceptable and meaningful functions, actually functionals, following the standard method in analysis. First, we have to define the most elementary and basic functions, namely polynomials in $\dot{B}(t)$'s. For example, we first discuss $(\dot{B}(t))^n$, in particular $(\frac{dB}{dt})^2$.

Now, recall the simplest case of the Itô formula $(dB(t))^2 = dt$. This implies $(\frac{dB}{dt})^2 = \frac{1}{dt}$, which has no meaning at all. Here we note that $(dB(t))^2$ is random, although it is infinitesimal. To define the square of $\dot{B}(t)$ we need some modification to overcome the difficulties above. Our idea is to apply the *renormalization* to the centered random quantity $(dB(t))^2 - dt$. Now we are ready to magnify $(dB(t))^2 - dt$ by multiplication as much as $(\frac{1}{dt})^2$. The renormalized square of $\dot{B}(t)$ is denoted by $: \dot{B}(t)^2 :$, a formal description of which is $\dot{B}(t)^2 - \frac{1}{dt}$.

Note To subtract off dt is simply an orthogonalization to the space H_1. Essential part of renormalization is *magnification* by $(\frac{1}{dt})^2$. Quite recently, we noticed a trick in the paper by Mikusinski [7], where the square of the delta function is discussed. We see some similarity, although he dealt with non random quantity.

To define a big space that includes $: \dot{B}(t)^2 :$ we may think of a generalization of what we did for $\dot{B}(t)$. For general functionals of $\dot{B}(t)$'s, generalization of the renormalization is too comlicated. We can, however, provide an efficient tool that can be applied to general case. It is the S-transform of white noise functional φ expressed in the form

$$S(\varphi(\dot{B}))(\xi) = \exp[-\frac{1}{2}\|\xi\|^2]E[e^{<\dot{B},\xi>}\varphi(\dot{B})].$$

If we apply the S-transform to an approximation $(\Delta B)^2$, we are given $\xi(t)^2\Delta^2 + \Delta$ plus higher order term. This tells us how to manage, i.e. renormalization to have acceptable quantity $\xi(t)^2$. This can be generalized, although we omit the details.

We are now ready to define a class of generalized white noise functionals in the general setup. There are two typical ways of introducing generalized white noise functionals rigourously. One is to use the Sobolev space structure (polynomials in $\dot{B}(t)$'s are managed specifically) and the other is an (infinite dimensional) analogue of the Schwartz space of the tempered distributions on R^d.

Use of the Sobolev Space Structure

It is well-known that the Hilbert space (L^2) of all functionals of Brownian motion with finite variance admits a direct sum decomposition, called Fock space:

$$(L^2) = \oplus_{n=0}^{\infty} H_n,$$

where H_n is the space of homogeneous chaos of degree n, or we understand that it is the space of the n-ple Wiener integrals in the Itô sense.

Generally, we know that there is a mapping that defines an isomorphic mapping from H_n to $\widehat{L^2(R^n)}$:

$$\pi_n : \ H_n \ \cong \ \widehat{L^2(R^n)},$$

where $\widehat{L^2(R^n)}$ is the subspace of $L^2(R^n)$ consisting of all symmetric functions. More precisely, for $\varphi \in H_n$ the following equality holds:

$$\|\varphi\| = \sqrt{n!}\|\pi_n\varphi\|_{L^2(R^n)}.$$

Our basic idea is to have a reasonable extension of the space H_n corresponding to a reasonable extension of $\widehat{L^2(R^n)}$.

A mathematically rigorous construction of the proposed space, obtained with the help of the Sobolev spaces, is as follows.

Denote by $K^m(R^n), m > 0$, the Sobolev space over R^n of order m. The subspace involving all symmetric functions of $K^m(R^n)$ will be denoted by $\widehat{K^m(R^n)}$. The space $\widehat{K^{-m}(R^n)}$ is the dual space of $\widehat{K^m(R^n)}$ with respect to the topology being equipped with $L^2(R^n)$. Thus we have another triple for every $m > 0$:

$$\widehat{K^m(R^n)} \subset \widehat{L^2(R^n)} \subset \widehat{K^{-m}(R^n)},$$

where the both inclusions are continuous injections.

Our favorable choice of the degree m of the Sobolev space is taken to be $(n+1)/2$. For one thing, members in $K^{(n+1)/2}(R^n)$ are continuous and the restriction of them to an $(n-1)$-dimensional hyperplane belongs to the space $\widehat{K^{n/2}(R^{n-1})}$, namely the degree decreases by $1/2$ when the restriction is made to a subspace of one dimension lower. This comes from the trace theorem in the theory of the Sobolev spaces. And so forth. These properties are convenient when the analysis on H_n is carried out, by taking the isomorphism between H_n which is the space of homogeneous chaos of degree n and $\widehat{L^2(R^n)}$ into account.

Let m be taken specially to be $(n+1)/2$ in the above triple involving symmetric Sobolev spaces. Then, we have

$$\widehat{K^{(n+1)/2}(R^n)} \subset \widehat{L^2(R^n)} \subset \widehat{K^{-(n+1)/2}(R^n)},$$

and we form

$$H_n^{(n)} \subset H_n \subset H_n^{(-n)}.$$

Each space of this triple is isomorphic to the corresponding symmetric Sobolev space. More precisely, we recall the isomorphism π_n, under which the H_n is isomorphic to $\widehat{L^2(R^n)}$, where the corresponding norms are different only by the constant $\sqrt{n!}$. Take the subspace $\widehat{K^{(n+1)/2}(R^n)}$ of $\widehat{L^2(R^n)}$, with a note that the former is equiped with a stronger topology. It is also noted that the isomorphism π_n can be restricted to any subspace, so that the restriction is denoted by the same symbol π_n. Let Q_n be the (continuous) injection such that

$$Q_n : \widehat{K^{(n+1)/2}(R^n)} \rightarrow \widehat{L^2(R^n)}.$$

Define

$$H^{(n)} = \pi_n^{-1} Q_n K^{\widehat{(n+1)/2}}(R^n).$$

The $H^{(n)}$ is topologized so as to be isomorphic to the space $K^{\widehat{(n+1)/2}}(R^n)$. The norms of these two spaces are different only by the constant $\sqrt{n!}$.

Based on the basic scalar product in H_n, the dual space of $H^{(n)}$ can be defined in the usual manner, and it is denoted by $H^{(-n)}$.

The norms in those spaces are denoted by $\|\cdot\|_n$, $\|\cdot\|$, and $\|\cdot\|_{-n}$, respectively. Thus, we have a Hilbert space $H_n^{(-n)}$ of *generalized white noise functionals* of degree n.

There remains a freedom on how to sum up the spaces $H_n^{(-n)}, n \geq 0$, with suitable weights. Choose an increasing sequence $c_n > 0$, and form a Hilbert space $(L^2)^+$ by the direct sum:

$$(L^2)^+ = \bigoplus c_n H_n^{(n)}, \tag{2}$$

where $\varphi \in (L^2)^+$ is of the form

$$\varphi = \sum_n \varphi_n, \quad \varphi_n \in H_n,$$

and where the $(L^2)^+$-norm $\|\cdot\|$ is defined by

$$\|\varphi\|_+^2 = \sum_n c_n^2 \|\varphi_n\|_m^2.$$

The direct sum given above forms a Hilbert space and its dual space $(L^2)^-$ is expressed in the form

$$(L^2)^- = \bigoplus c_n^{-1} H_n^{(-n)}. \tag{3}$$

The definitions of $(L^2)^+$ and $(L^2)^-$ given above have rather formal significance, but they can show our idea behind the formality. The correct meaning is as follows.

The weighted algebraic sum $\sum c_n H_n^{(n)}$ is given without any difficulty. The topology is defined by the semi-norm $\|\cdot\|_+$. Set $(L^2)^+$ be the completion of the vector space spanned by those φ's that have finite seminorm. Eventually $\|\cdot\|_+$ becomes a norm. It is easy to see that $(L^2)^+$ is a separable Hilbert space with the Hilbertian norm $\|\cdot\|_+$.

Based on the (L^2)-norm we can introduce a dual space $(L^2)^-$ of the Hilbert space $(L^2)^+$. Thus, each space $H_n^{(-n)}$ may be viewed as a subspace of $(L^2)^-$ and a member ψ of $(L^2)^-$ may be expressed as a sum of the form $\sum \psi_n, \psi_n \in H_n^{(-n)}$. We claim that $(L^2)^-$ is a Hilbert space with the norm $\|\cdot\|_-$ defined by $\|\psi\|_-^2 = \sum c_n^{-2}\|\psi_n\|_{-n}^2$.

Naturally, we are given a triple

$$(L^2)^+ \subset (L^2) \subset (L^2)^-.$$

The space $(L^2)^+$ consists of test functionals, while $(L^2)^-$ is the space of *generalized white noise functionals.*

The S-transform can generalize the homeomorphism to establish the space F^-, which is to be the extension of F in such a way that

$$(L^2)^- \sim F^{-1} \tag{4}$$

Examples of $(L^2)^-$-functionals are given in the following.

Example 1. White noise $\dot{B}(t)$, and the renormalized square : $\dot{B}(t)^2 := \dot{B}(t)^2 - \frac{1}{dt}$ where $S(:\dot{B}(t)^2:)$ is equal to $\xi(t)^2$.

Example 2. The Wick powers of $\dot{B}(t)$. We choose : $\dot{B}(t)^n$: . It belongs to $H_n^{(-n)}$. The associated S-transform is $\xi(t)^n$ and the kernel function is $\delta_t^{n\otimes}$. As for the definitions of the Wick product we can see the book by Hida-Kuo-Potthoff-Streit.

Example 3. $\delta_0(B(t))$ (the Donsker's delta function) is a member of $(L^2)^-$. Its S-transform is expressed simply in the form:

$$S(\delta_a(B(t)))(\xi) = \frac{1}{\sqrt{2\pi t}} \exp[-\frac{1}{2t}(a - \int_0^t \xi(u)du)^2].$$

This equation comes from the well-known formula of a delta function, namely

$$\delta_0(u) = \frac{1}{2\pi}\int e^{itu}dt.$$

Example 4. $\varphi_c(\dot{B}) = N \exp[c \int \dot{B}(t)^2 dt]$, $c \neq 1/2$, N being a normalizing constant. It is a generalized white noise functional that plays a role of the Gauss kernel in $(L^2)^-$. The S-transform of φ_c is given by

$$(S\varphi_c)(\xi) = \exp\left[\frac{c}{1-2c}\|\xi\|^2\right].$$

In reality the (multiplicative) renormalizing constant N is determined so as the S-transform is given by the above acceptable formula.

An Analogue of the Schwartz Space

The second method of defining a space of generalized white noise functionals is using an infinite dimensional analogue of the Schwartz space over R^1.

By analogy with

$$S \subset L^2(R) \subset S'$$

we have

$$(S) \subset (L^2) \subset (S)^*.$$

The $(S)^*$ is the space of generalized white noise functionals. The idea of the construction of the space is the second quantization of the operator

$$A = -\frac{d^2}{du^2} + u^2 + 1.$$

Two spaces $(L^2)^-$ and $(S)^*$ are slightly different and they enjoy their own characteristics. One big advantage of $(S)^*$ is that there is a characterization theorem of members in $(S)^*$ due to Potthoff and Streit.

Theorem 1 (Potthoff and Streit) A functional $U(\xi)$ is an S-transform of a functional in $(S)^*$ if and only if

1) U is ray analytic,

2) there exists $p \in N$ and $c_1, c_2 > 0$ such that for all $z \in \mathbf{C}, \xi \in E$, the following inequality holds:

$$|U(\xi)| \leq c_1 \exp[c_2|z|^2\|\xi\|_p^2],$$

where $\|\cdot\|$ is one of the sequence of norms defining the topology of E.

INFINITE DIMENSIONAL SPACES

Coming back to the space $H_1^{(-1)}$ spanned by the variables $\dot{B}(t)$'s that we have established by the step of Reduction in the beginning of the last section.

The cardinal number of variables $\dot{B}(t)$'s is continuum. They are independent and even seems to be linearly independent. There arises a question if the space $H_1^{(-1)}$ is *non-separable*, since there are linearly independent vectors as many as continuum. On the other hand, all those basis vectors come from a Brownian motion, which spans a separable space. Hence, we may have a contradiction.

The above conclusion is obviously *wrong*. We have to remind the definition of *linearly independent system*. It is true that if for different t_j's (finitely many)

$$\sum_1^n a_j \dot{B}(t_j) = 0,$$

then all a_j's are equal to 0. This is equivalent to the fact that if

$$\sum_1^n a_j \delta_{t_j} = 0$$

then all a_j's are equal to 0.

We understand that any finite collection of $\dot{B}(t)$'s forms a linearly independent system in $H_1^{(-1)}$. So the dimension of $H_1^{(-1)}$ is infinite (at least countably infinite). The set of collections is a (semi-)ordered set and we may consider lattice sum to have maximal. Unfortunately the Zorn's Lemma cannot be applied.

There is an obvious assertion:

Proposition 1 *A linear combination* $\sum_j c_j \dot{B}(t_j)$ *for different* t_j*'s is a vector in* $H_1^{(-1)}$ *if and only if* $\sum_j |c_j|^2 < \infty$.

There is an obvious remark. When linearly independence is discussed, a linear combination is either finite or at least the sum should belong to the space in question.

By Proposition 1, the space $H_1^{(-1)}$ is *at least countably infinite dimensional.* Ovbiously it is not continuously infinite dimensional, so that we can simply say it is *countably infinite dimensional.*

Theorem 2 Every space $H_n^{(-n)}$ is countably infinite dimensional. So is $(L^2)^-$.

Proof. $H_n^{(-n)}$ is the symmetric tensor product of $H_1^{(-1)}$, which implies the assertion. The direct sum of them, that is $(L^2)^-$, is the same.

Another concept for a collecton of basic vectors is the totality. Let us remind the definition to claim that a system of vectors is total. A collection of vectors is said to be *total* if the members of the collection span a dense linear subspace.

Proposition 2 *Let the set* $A = \{t_j; j \in N\}$, *where* N *is the set of natural numbers, be dense in* R^1. *Then, the collection* $\{\dot{B}(t_j), j \in A\}$ *is total in* H_1^{-1}.

Proof. The delta function δ_t is associated to $\dot{B}(t)$. The Fourier transform of the delta functio δ_t is $\frac{1}{\sqrt{2\pi}} e^{iyt}$. It belongs to $K^{-1}(R^1)$ and the collection $\{\frac{e^{iyt_j}}{\sqrt{1+y^2}}; j \in A\}$ spans the whole space $K^{-1}(R^1)$. Hence we prove $\dot{B}(t_j)$'s span the space $H_1^{(-1)}$, i.e. total.

REFERENCES

1. T. Hida, "Analysis of Brownian Functionals," *Carleton Math. Lecture Notes. no. 13*, July, 1975.

2. T. Hida and Si Si, *Lectures on White Noise Functionals*, World Scientific, Singapore, 2008.

3. T. Hida, "The concept of infinite dimension in the analysis of white noise generalized functionals," *RIMS Ojima Conf.* Oct. 2007. (in Japanese)

4. P. Lévy, *Processus stochastiques et mouvement brownien*, Gauthier-Villars, 1948, 2éme ed. 1965.

5. P. Lévy, *Problèmes concrets d'analyse fonctionnelle*, Gauthier-Villars, 1951.

6. J. -L. Lions, "The earth, planet the role of mathematics and of super computers". (in Spanish), *Inst. Espana. Espasa Calpe.* 1990.

7. J. Mikusinski, "On the square of the Dirac Delta-distribution" *Bull. de l'Academie Polonaise des Sciences. Ser. Sci. math. astr. et Phys.* XIV, no. **9**, 511-513 (1966).

8. Si Si, An aspect of quadratic Hida distributions in the realization of a duality between Gaussian and Poisson noises. *Infinite Dim. Analysis, Quantum Prob. and Related Topics* **11**, to appear.

9. L. Streit, "Feynman Integrals as Generalized Functions on Path Space: Things done and Open Problems," in *Proceedings of the International Conference on Path Integrals - New Trends and Perspectives*, edited by W. Janke and A. Pelster, World Scientific, Singapore, 2008, to appear.

Selecta from a Life-Long Obsession with Path Integrals

John R. Klauder

Department of Physics and Department of Mathematics
University of Florida
Gainesville, FL 32611

Abstract. The definition and interpretation of canonical, phase space path integrals has evolved over many years to achieve a form that now admits a correct and rigorous formulation, which is also covariant under canonical coordinate transformations. Such formulations involve coherent state representations, which, in their modern version, were originally introduced as an alternative tool to construct phase space path integrals. Moreover, coherent state representations lead to physical interpretations that are more natural than those afforded by more traditional representations. Suitable continuous time regularization procedures lead to a covariant phase space path integral formulation that greatly clarifies the vague phrase that canonical quantization requires Cartesian coordinates.

Keywords: Coherent state path integrals, Wiener measure regularization, canonical coordinate transformations
PACS: 01.30.Cc, 02.30.Cj, 02.50.Ey, 03.65.Db

INTRODUCTION

Path integrals for quantum mechanics were introduced by Feynman in his 1942 thesis, and first published in 1948 [1]. The present author became acquainted with the subject in 1956 through the lectures of John A. Wheeler, who was Feynman's thesis advisor (and later the author's advisor as well).

The original path integral was of the form

$$K = \mathcal{N} \int e^{(i/\hbar)\int[\frac{1}{2}\dot{x}^2 - V(x)]\,dt}\, \mathscr{D}x\,, \tag{1}$$

for a time interval $0 \leq t \leq T$, $T > 0$, integrated over all paths that start at $x(0) = x'$ and end at $x(T) = x''$ for suitable potentials V, where $\mathscr{D}x$ denotes a putative translationally invariant measure. The factor $\mathcal{N}$ is an overall constant that may be fixed later by other means. The resultant function $K = K(x'', T; x', 0)$, where

$$K(x'', T; x', 0) = \langle x''|e^{-i\mathscr{H}T/\hbar}|x'\rangle\,, \tag{2}$$

$\mathscr{H} = \frac{1}{2}P^2 + V(Q)$ is the Hamiltonian operator, and $Q|x\rangle = x|x\rangle$, which indicates that the states $|x\rangle$ are sharp eigenvectors of the position operator Q. The K function is the propagator for solutions to the associated Schrödinger equation. As it stands, this path integral applies to many problems, but it is apparently limited to those examples whose classical action is given by

$$I = \int[\tfrac{1}{2}\dot{x}^2 - V(x)]\,dt\,, \tag{3}$$

CP1021, *Stochastic and Quantum Dynamics of Biomolecular Systems*,
Proceedings of the 5th Jagna International Workshop, edited by C. C. Bernido and M. V. Carpio-Bernido

which by no means covers all potentially interesting systems.

The limitation to actions of the form (3) is overcome with the *phase space path integral*, introduced by Feynman in 1951 [2], which is given by

$$K = \mathcal{M} \int e^{(i/\hbar)\int [p\dot{q} - H(p,q)]\,dt}\, \mathcal{D}p\,\mathcal{D}q\,. \tag{4}$$

Here, we see the classical action

$$I = \int [p\dot{q} - H(p,q)]\,dt \tag{5}$$

involving phase space paths, $p(t)$ and $q(t)$, over a similar time interval, $0 \le t \le T$, $T > 0$, where $H(p,q)$ is the Hamiltonian. Some limitation on the range of these paths should also be part of the definition of the integral. The original form of this integral [2] involved integration over all paths $p(t)$ as well as all paths $q(t)$ which satisfy $q(0) = q'$ and $q(T) = q''$, leading to the propagator $K = K(q'', T; q', 0)) = \langle q''| e^{-i\mathcal{H}T/\hbar} |q'\rangle$, which again is the propagator for solutions of the Schrödinger equation for the quantum system under consideration. Indeed, if

$$H(p,q) = \tfrac{1}{2}p^2 + V(q)\,, \tag{6}$$

an analogue of a Gaussian integral leads back to the original path integral form given in (1). (As a reminder of which path integral is being considered, the author prefers to use x as the coordinate for the configuration space path integral and to use q as the coordinate for the phase space path integral.)

At first glance, formal rules of integration would seem to be adequate; however, a closer look shows how wrong this view would be. For one thing, nothing has been said about which choice of phase space coordinates are to be used in this evaluation. A casual observer would say it shouldn't matter which choice of coordinates are used because the integral appears to be covariant under changes of canonical coordinates. Such covariance is clear for the classical action and it should apply as well to the measure $\mathcal{D}p\,\mathcal{D}q$ thanks to the Liouville Theorem regarding the canonical invariance of the phase space measure $dp\,dq$. On the other hand, if covariance under canonical coordinate transformations actually held, it would imply that dissimilar classical systems (free particle, harmonic oscillator, anharmonic oscillator, etc.) all had identical quantum spectra, which is well known to be false. Thus, despite appearances, we are forced to conclude that the path integral for quantum mechanics, (1) or (4), is *not* well defined. This lack of definition is both a mathematical and a physical deficiency. In point of fact, the formal path integral is at best merely "semi-classical" in character. This is not a new idea, but it generally is not given its proper importance when many investigators evaluate, even approximately, such expressions. In the author's view, providing a proper path integral definition is surely the **first issue**, and in many ways the **main issue**, in any attempt to evaluate path integrals! And, finding proper definitions for formal path integrals provides the common link among the several procedures that are discussed subsequently in this paper.

BASIC RULES

Dirac has emphasized [3] that the transformation of a classical Hamiltonian $H(p,q)$ into the correct quantum Hamiltonian operator $\mathscr{H}(P,Q)$ involves the use of "Cartesian coordinates", and not more general curvilinear coordinates, when making the traditional substitution $p \to P$, $q \to Q$, where $[Q,P] = i\hbar$; additional discussion of the phrase "Cartesian coordinates" is given below.[1] Path integrals cannot avoid that dictum as well. Thus, when natural lattice regularizations are introduced, it is *implicit* that the chosen coordinates are Cartesian. We have in mind the lattice prescription for (1) given by

$$K = \lim_{N\to\infty} A_N \int \cdots \int \exp\{(i/\hbar)\Sigma_{l=0}^{N}[(x_{l+1}-x_l)^2/2\varepsilon - \varepsilon V(x_l)]\}\,\Pi_{l=1}^{N}\,dx_l\,, \tag{7}$$

where $x'' = x_{N+1}$, $x' = x_0$, and $(N+1)\varepsilon = T$. The normalization factor $A_N = (2\pi i\hbar\varepsilon)^{-(N+1)/2}$. Likewise, the lattice prescription for (4) may be given by

$$K = \lim_{N\to\infty} \int \cdots \int \exp\{(i/\hbar)\Sigma_{l=0}^{N}[p_{l+1/2}(q_{l+1}-q_l) \\ -\varepsilon H(p_{l+1/2},(q_{l+1}+q_l)/2)]\}\,\Pi_{l=0}^{N}dp_{l+1/2}/(2\pi\hbar)\,\Pi_{l=1}^{N}dq_l\,, \tag{8}$$

where all p variables are integrated, and where $q'' = q_{N+1}$, $q' = q_0$, and $(N+1)\varepsilon = T$, as before. The notation $p_{l+1/2}$ signifies that the p variables are evaluated at (say) mid-lattice intervals because the meaning of q and the meaning of p is that of *sharp eigenvalues*, i.e., $Q|q\rangle = q|q\rangle$ and $P|p\rangle = p|p\rangle$, and therefore they cannot be defined at the same time. Observe that an expression such as (8), which is basically well defined, shows that the number of p and q integrations are *not* the same, a conclusion that also follows immediately from the general combination formula given by

$$K(q'',T+S;q',0) = \int K(q'',T+S;q,S)\,K(q,S;,q',0)\,dq\,; \tag{9}$$

this expression is also useful to help determine the overall normalization of the propagator.

While (8) is mathematically correct, it is physically unnatural. This situation arises because the interpretation of (8) involves the classical action being evaluated, at alternate lattice times, at sharp q, then at sharp p, then at sharp q, then at sharp p, etc. One cannot expect that true classical paths could arise from such an interpretation, because that would require p and q to be defined at the same time. By analogy, recall that epicycles provided a mathematically correct characterization of planetary motion, but it was physically unnatural.

[1] The errors that can arise when using non-Cartesian coordinates are nicely illustrated with a simple two-dimensional example in Shankar's quantum mechanics book [4].

COHERENT STATE PATH INTEGRALS

Alternate use of the resolutions of unity afforded by the sharp Q states $|q\rangle$, given by $\int |q\rangle\langle q|\,dq = 1$, as well as that afforded by the sharp P states, given by $\int |p\rangle\langle p|\,dp = 1$, plays an important role in deriving the phase space form of the path integral in (8); that alternate use of distinct resolutions of unity also gives rise to the unnatural, alternately sharp p, sharp q, ..., description of the "paths" involved. It is the introduction of a single, universally suitable resolution of unity based on coherent states that leads to a more natural formulation.

Coherent states are not eigenvectors of any Hermitian operator. Canonical coherent states may be defined by

$$|p,q\rangle \equiv e^{-iqP/\hbar}\, e^{iqP/\hbar}\,|\eta\rangle\,, \qquad (p,q)\in\mathbb{R}^2\,, \tag{10}$$

where $|\eta\rangle$ denotes a normalized fiducial vector. Oftentimes, it is useful to choose $|\eta\rangle = |0\rangle$, where $(Q+iP)\,|0\rangle = 0$, namely, $|0\rangle$ is the ground state of an harmonic oscillator with unit mass and unit angular frequency. In any case, for any choice of $|\eta\rangle$, it follows [5] that

$$1 = \int |p,q\rangle\langle p,q|\,d\mu(p,q)\,, \tag{11}$$

where $d\mu(p,q) = dp\,dq/(2\pi\hbar)$ and the integration runs over the whole phase space plane. Unlike the resolutions afforded by the states $\{|q\rangle\}$ (or $\{|p\rangle\}$, that afforded by the coherent states involves an *overcomplete* set of states $\{|p,q\rangle\}$.

If we impose the modest conditions of "physical centrality" given by $\langle\eta|P|\eta\rangle = \langle\eta|Q|\eta\rangle = 0$, which is automatically fulfilled when $|\eta\rangle = |0\rangle$, then it follows that

$$\begin{aligned}\langle p,q|P|p,q\rangle &= p\,,\\ \langle p,q|Q|p,q\rangle &= q\,,\end{aligned} \tag{12}$$

which implies that the physical meaning of p and q is that of *mean* values rather than sharp eigenvalues. As such it is clear that we can specify the mean values of P and Q simultaneously. For the sake of convenience in what follows, we assume hereafter that $|\eta\rangle = |0\rangle$, an oscillator ground state.

Starting once again with the formal phase space path integral (4), we choose an alternative regularization to give it meaning, namely,

$$\begin{aligned}K = \lim_{N\to\infty}\int\cdots\int \exp\{&\Sigma_{l=0}^{N}[(i/2\hbar)(p_{l+1}+p_l)(q_{l+1}-q_l)\\ &-(1/4\hbar)[(p_{l+1}-p_l)^2+(q_{l+1}-q_l)^2]\\ &-i(\varepsilon/\hbar)H(p_{l+1},q_{l+1};p_l,q_l)]\}\,\Pi_{l=1}^{N}d\mu(p_l,q_l)\,.\end{aligned} \tag{13}$$

Here, we have introduced

$$H(p_{l+1},q_{l+1};p_l,q_l) \equiv \langle p_{l+1},q_{l+1}|\mathscr{H}(P,Q)|p_l,q_l\rangle/\langle p_{l+1},q_{l+1}|p_l,q_l\rangle\,. \tag{14}$$

In this case, $K = K(p'',q'',T;p',q',0)$ where $p_{N+1},q_{N+1} = p'',q''$, and $p_0,q_0 = p',q'$, and as before $(N+1)\varepsilon = T$. Moreover, it follows that

$$K(p'',q'',T;p',q',0) = \langle p'',q''|e^{-i\mathscr{H}T/\hbar}|p',q'\rangle\,. \tag{15}$$

This function is also a propagator for solutions to the Schrödinger equation but in a different representation. First, recall that $\psi(q) = \langle q|\psi\rangle$ yields the Schrödinger representation of the abstract vector $|\psi\rangle$ with square norm given by

$$\| |\psi\rangle \|^2 = \int \langle\psi|q\rangle\langle q|\psi\rangle\, dq = \int |\psi(q)|^2\, dq \,. \tag{16}$$

In like manner, $\psi(p,q) = \langle p,q|\psi\rangle$ yields the coherent state representation of the abstract vector $|\psi\rangle$ with square norm being given by

$$\| |\psi\rangle \|^2 = \int \langle\psi|p,q\rangle\langle p,q|\psi\rangle\, d\mu(p,q) = \int |\psi(p,q)|^2\, d\mu(p,q) \,. \tag{17}$$

The interpretation of $\psi(q)$ is the probability amplitude (density) for the particle to be found at q; the interpretation of $\psi(p,q)$ is the probability amplitude for the state of the system $|\psi\rangle$ to be found in the coherent state $|p,q\rangle$. While the $|q\rangle$ states are orthogonal for different q values, that is not the case for different arguments of the coherent states.

The abstract evolution of $|\psi,0\rangle$, the state vector at time $t = 0$, is given as usual by $|\psi,T\rangle = \exp[-i\mathscr{H}T/\hbar]\,|\psi,0\rangle$. In the Schrödinger representation this equation reads

$$\psi(q'',T) = \langle q''|e^{-i\mathscr{H}T/\hbar}|\psi,0\rangle = \int K(q'',T;q,0)\,\psi(q,0)\,dq \,. \tag{18}$$

In like manner, in the coherent state representation, we find

$$\begin{aligned}\psi(p,''q'',T) &= \langle p'',q''|e^{-i\mathscr{H}T/\hbar}|\psi,0\rangle \\ &= \int K(p'',q'',T;p,q,0)\,\psi(p,q,0)\,d\mu(p,q) \,,\end{aligned} \tag{19}$$

which has a similar mathematical appearance.

Finally, we observe that (13) is both mathematically correct and physically natural. The latter comment arises because the variables p and q are physically interpreted as mean values and thus may be simultaneously specified, as indeed they are in (13). Moreover, there are the same number of p and q integrations involved. This feature is compatible with the new form of the combination formula given by

$$\begin{aligned}&K(p'',q'',T+S;p',q',0) \\ &\quad = \int K(p'',q'',T+S;p,q,S)\,K(p,q,S;,p',q',0)\,d\mu(p,q) \,.\end{aligned} \tag{20}$$

A PHASE SPACE METRIC

In classical mechanics, the phase space M has a symplectic form ω and carries no metric. This is mathematically natural since $M = T^*Q$, namely, it involves the dual to the tangent space of the configuration space Q. The symplectic form is invariant under the wide class of canonical coordinate transformations given by $\overline{p} = \overline{p}(p,q)$ and $\overline{q} = \overline{q}(p,q)$, where, e.g.,

$$\overline{p}\,d\overline{q} = p\,dq + dF(\overline{q},q) = p\,dq + dG(p,q) \,. \tag{21}$$

An exterior derivative of (21) leads to $d\overline{p}\wedge d\overline{q} = dp\wedge dq$, which is just Liouville's Theorem.

While there is no phase space metric in classical mechanics, we can see a hint of one in quantum mechanics if we look closely at (13) and focus on the factor

$$(p_{l+1}-p_l)^2+(q_{l+1}-q_l)^2 \,. \tag{22}$$

This factor does not survive in the classical action for continuous and differentiable paths, but it arises from the coherent state overlap function and is an important part of the overall integral. Let us trace out the appearance of a metric as induced by the coherent states themselves.

Let $d|p,q\rangle \equiv |p+dp,q+dq\rangle - |p,q\rangle$. It then follows from (10), with $|\eta\rangle = |0\rangle$, that

$$\theta(p,q) \equiv i\hbar\,\langle p,q|d|p,q\rangle = p\,dq\,, \tag{23}$$

as well as

$$d\sigma(p,q)^2 \equiv 2\hbar\,[\,\|d|p,q\rangle\,\|^2 - |\langle p,q|d|p,q\rangle|^2\,] = dp^2+dq^2\,. \tag{24}$$

Observe, that the metric $d\sigma^2$ decribes a two-dimensional flat space in Cartesian coordinates. If we describe the coherent states in terms of different canonical coordinates by

$$|\overline{p},\overline{q}\rangle \equiv |p(\overline{p},\overline{q}),q(\overline{p},\overline{q})\rangle = |p,q\rangle\,, \tag{25}$$

it follows that

$$i\hbar\,\langle \overline{p},\overline{q}|d|\overline{p},\overline{q}\rangle = \overline{p}\,d\overline{q} + d\overline{G}(\overline{p},\overline{q})\,, \tag{26}$$

$$d\overline{\sigma}(\overline{p},\overline{q})^2 = A(\overline{p},\overline{q})\,d\overline{p}^2 + 2B(\overline{p},\overline{q})\,d\overline{p}\,d\overline{q} + C(\overline{p},\overline{q})\,d\overline{q}^2\,, \tag{27}$$

both of which transform covariantly under canonical coordinate transformations, with the metric still describing a flat space, but, generally, now in curvilinear coordinates.

How can a phase space metric be useful? Consider the question "What physical meaning is to be given to the classical Hamiltonian described in suitable canonical coordinates by the expression $\tilde{p}$". There is no 'label' attached to the expression $\tilde{p}$ that explains its physical significance; all that is offered is just the expression $\tilde{p}$. Clearly, there is not enough information to determine the physical meaning to be attached to the symbol $\tilde{p}$, but a metric can be helpful in this regard. Suppose we assign a flat geometry to the phase space and introduce a set of Cartesian coordinates p and q so that the metric for this flat space reads $d\sigma^2 = dp^2+dq^2$. In such coordinates we declare that the Hamiltonian in question is described by $\frac{1}{2}(p^2+q^2)$, and represents a certain harmonic oscillator; this assignment is implicitly carried out for all other physical systems of possible interest in this coordinate system as well. Next, we allow canonical coordinate transformations, e.g.,

$$\tilde{p} = \tilde{p}(p,q) = \tfrac{1}{2}(p^2+q^2) \tag{28}$$

$$\tilde{q} = \tilde{q}(p,q) = \tan^{-1}(q/p)\,, \tag{29}$$

in which the expression for the oscillator Hamiltonian becomes

$$\tilde{p} = \tilde{p}(p,q) = \tfrac{1}{2}(p^2+q^2)\,. \tag{30}$$

The physical meaning of the Hamiltonian in the new coordinates is not lost but is encoded in the form taken by the metric under the same transformation, namely, by the fact that now $d\sigma^2$ reads

$$d\tilde{\sigma}^2 = (2\tilde{p})^{-1}\, d\tilde{p}^2 + (2\tilde{p})\, d\tilde{q}^2 \,. \tag{31}$$

This form implies that to find the correct physical meaning of $\tilde{p}$ it is first necessary to recast $\tilde{p}$ into coordinates in which the metric is Cartesian; in short, the metric keeps track of the physical meaning of the mathematical symbols that describe various quantities, such as the Hamiltonian.

WIENER MEASURE REGULARIZATION

The previous section taught us that adding a metric to a flat classical phase space allows us to keep track of the physical meaning of mathematical expressions for the Hamiltonian. In this section we again start with (4) for the formal phase space path integral but give it meaning by introducing another form of regularization different from the lattice regularization in either (8) or (13). In particular, we consider

$$K = \lim_{\nu\to\infty} \mathscr{M}_\nu \int e^{(i/\hbar)\int[p\dot{q}-H(p,q)]\,dt}\, e^{-(1/2\nu)\int(\dot{p}^2+\dot{q}^2)\,dt}\, \mathscr{D}p\, \mathscr{D}q \,, \tag{32}$$

where we have introduced a damping factor that formally goes to unity in the limit that $\nu \to \infty$. This expression can be reexpressed rigorously in the form

$$K = \lim_{\nu\to\infty} (2\pi\hbar)\, e^{\nu T/2\hbar} \int e^{(i/\hbar)\int[p\,dq - H(p,q)\,dt]}\, d\mu_W^\nu(p,q) \,, \tag{33}$$

where $\mu_W^\nu(p,q)$ denotes a pinned Wiener measure with diffusion constant ν, fixed so that $p(T), q(T) = p'', q''$ and $p(0), q(0) = p', q'$, and which is normalized so that

$$\int d\mu_W^\nu(p,q) = (2\pi\nu T)^{-1}\, e^{-(1/2\nu T)[(p''-p')^2+(q''-q')^2]} \,. \tag{34}$$

Moreover, the integral $\int p\,dq$ denotes a Stratonovich, i.e., mid-point rule, stochastic integral.

For each $\nu < \infty$, note that (33) is a rigorously defined path integral, concentrated on continuous (but nowhere differentiable) phase space paths. In the present case, $K = K(p'',q'',T;p',q',0)$, and again

$$K(p'',q'',T;p',q',0) = \langle p'',q''|\, e^{-i\mathscr{H}T/\hbar}\, |p',q'\rangle \,, \tag{35}$$

namely, the propagator expressed in the coherent state representation. Since the integral expression (33) is well defined, it follows that the Wiener measure regularization – like the lattice space regularization before it – has picked its own choice of operator Hamiltonian including any possible factor ordering ambiguity. In particular, it follows that

$$\mathscr{H}(P,Q) = \int H(p,q)\, |p,q\rangle\langle p,q|\, d\mu(p,q) \,, \tag{36}$$

an expression equivalent to the statement that the Hamiltonian is *antinormal ordered.*

It is especially noteworthy that by inserting the Wiener measure regularization we have also introduced into (32) the very metric dp^2+dq^2 that was found useful to monitor the physical meaning of the mathematical expression for the Hamiltonian. As a well defined integral, (33) permits a change of integration variables such as that given by a canonical coordinate transformation. Under such a transformation from (p,q) to $(\overline{p},\overline{q})$, (33) becomes

$$\overline{K}(\overline{p}'',\overline{q}'',T;\overline{p}',\overline{q}',0) \\ = \lim_{\nu\to\infty}(2\pi\hbar)\,e^{\nu T/2\hbar}\int e^{(i/\hbar)\int[\overline{p}\,d\overline{q}+d\overline{G}(\overline{p},\overline{q})-\overline{H}(\overline{p},\overline{q})\,dt]}\,d\overline{\mu}^{\nu}_{W}(\overline{p},\overline{q})\,, \tag{37}$$

which according to (25) asserts that

$$\overline{K}(\overline{p}'',\overline{q}'',T;\overline{p}',\overline{q}',0) = \langle\overline{p}'',\overline{q}''|e^{-i\mathcal{H}T/\hbar}|\overline{p}',\overline{q}'\rangle\,. \tag{38}$$

In the present case, the regularization still involves two-dimensional Brownian motion on the plane, but now described in general with curvilinear coordinates. A change in the phase space metric, such as illustrated in (27), encodes the change in the mathematical description of the Hamiltonian so as to maintain its original physical meaning. This connection is also evident in the integral relation

$$\mathcal{H}(P,Q) = \int\overline{H}(\overline{p},\overline{q})\,|\overline{p},\overline{q}\rangle\langle\overline{p},\overline{q}|\,d\mu(\overline{p},\overline{q}) \tag{39}$$

obtained from (36) by a well-defined canonical coordinate transformation in which we note that the basic Hamiltonian operator remains unchanged.

It is clear that the continuous time regularization offered by (33) leads to a quantization procedure that is functional, rigorous, covariant under canonical coordinate transformations, and also correct in that it agrees with experiment (since it agrees with Schrödinger's formulation, and that already agrees with experiment). Other formulations that are covariant under canonical coordinate transformations, such as Geometric Quantization, do not agree with traditional quantum mechanics and thus are not correct. The "cost" of obtaining the correct, covariant, functional version of canonical quantum mechanics reviewed in this paper has been the introduction of a metric on classical phase space and using that metric in supporting Brownian motion on the phase space. Even for a single p and q, as we have been discussing, we require a two-dimensional manifold which we have chosen to be flat. It may be shown that a flat space is a direct consequence of using the Heisenberg operator variables, or rather the Weyl group, to build our coherent states; the flat space is not dependent on the choice of $|0\rangle$ as the fiducial vector. However, having been led to a flat space, it follows that there are a select subset of coordinate systems that are Cartesian, and it is in just those coordinate systems that the algebraic form of the operator Hamiltonian expressed in the variables P and Q is identical to order $\hbar^0$ with the algebraic form of the classical Hamiltonian expressed in the variables p and q. This is one way of interpreting Dirac's dictum that the classical Hamiltonian must be expressed in Cartesian coordinates; indeed, the way we have described matters clarifies and extends the meaning of Dirac in that Cartesian coordinates even apply to a single p and a single q for which Cartesian has no meaning when applied to each variable separately, but can and does apply to both variables taken together. Moreover, it is inadequate to claim that Cartesian coordinates should only apply to higher dimensional $\{p_j\}$ and $\{q_k\}$ variables separately, for that would imply that a metric of the form (summation understood) $dp_j^2 + 2f_{jk}(p,q)\, dp_j\, dq_k + dq_k^2$, for general functions $\{f_{jk}(p,q)\}$ which lead to a positive definite metric, would imply a correct quantization. This is undoubtedly false. The only correct conclusion is that the phase space coordinates as a whole must be Cartesian, and this is the proper way that the Dirac dictum should be interpreted. Because coordinates and momenta are normally different, rather like "apples" and "oranges", the need to put them together in a phase space metric, requires going beyond the level of classical mechanics. Only the introduction of the Wiener measure regularization involving a flat phase space, as we have done – or some other procedure that is equivalent to the one we have introduced – leads to a suitable, covariant canonical quantization procedure.[2]

HISTORICAL REMARKS

Canonical coherent states were introduced in [6] and applied to the construction of phase space path integrals. The general theory of coherent states for arbitrary fiducial vectors was carefully developed in [5]. A precursor to Wiener regularization appeared in [7], while the full story arrived in [8, 9]. An appreciation of the covariance and geometrical implications appeared in [10], and the primary role of the phase space metric was emphasized in [11]. Finally, the history of Wiener measure regularizations for configuration space and phase space path integrals was summarized in [12].

ACKNOWLEDGMENTS

It is a pleasure to thank Profs. C. Bernido and M.V. Bernido for their splendid and gracious hospitality during the occasion of the 5th International Jagna Workshop held in Jagna, Bohol, The Philippines, in January, 2008. Congratulations are offered to Profs. L. Streit and F. Wiegel whose 70th birthdays were celebrated at this workshop.

REFERENCES

1. R.P. Feynman, "Space-time Approach to Non-relativistic Quantum Mechanics", Rev. Mod. Phys. **20**, 367–387 (1948).
2. R.P. Feynman, "An Operator Calculus having Application in Quantum Electrodynamics", Phys. Rev. **84**, 108–128 (1951).
3. P.A.M. Dirac, *The Principles of Quantum Mechanics*, (Oxford Press, Oxford, 1947), 3rd Ed., p. 114 (footnote).

---

[2] Although not the subject of this review, it is noteworthy that the use of a phase space metric for a suitable spherical surface or a pseudospherical surface, leads to a correct, but different and noncanonical, covariant quantization procedure for alternative kinematics, e.g., for generators of the group SU(2) or SU(1,1), in the cases mentioned in this footnote.

4. R. Shankar, *Principles of Quantum Mechanics*, (Plenum Press, New York, 1988), Exercise 7.4.10, p. 225.
5. J. McKenna and J.R. Klauder, "Continuous-Representation Theory IV. Structure of a Class of Function Spaces Arising from Quantum Mechanics", J. Math. Phys. **5**, 878–896 (1964).
6. J.R. Klauder, "The Action Option and the Feynman Quantization of Spinor Fields in Terms of Ordinary *c*-Numbers", Annals of Physics **11**, 123–168 (1960).
7. J.R. Klauder, "Path Integrals for Affine Variables", in Functional Integration, eds. J.-P. Antoine and E. Tirapegui, pp. 101–119 (Plenum Pub. Corp., 1980).
8. I. Daubechies and J.R. Klauder, "Quantum Mechanical Path Integrals with Wiener Measures for All Polynomial Hamiltonians, II", J. Math. Phys. **26**, 2239–2256 (1985).
9. I. Daubechies, J.R. Klauder, and T. Paul, "Wiener Measures for Path Integrals with Affine Kinematic Variables", J. Math. Phys. **28**, 85–102 (1987).
10. J.R. Klauder, "Quantization *Is* Geometry, After All", Annals of Physics **188**, 120–141 (1988).
11. J.R. Klauder, "Metrical Quantization", in *Quantum Future*, eds. P. Blanchard and A. Jadczyk, (Springer-Verlag, Berlin, 1999), pp. 129–138.
12. J.R. Klauder, "The Feynman Path Integral: An Historical Slice", in *A Garden of Quanta*, Eds. J. Arafune, et al (World Scientific, Singapore, 2003), pp. 55–76.

The Feynman integrand as a white noise distribution beyond perturbation theory

Martin Grothaus and Anna Vogel

Mathematics Department, University of Kaiserslautern, P.O.Box 3049, 67653 Kaiserslautern, Germany

Abstract. In this note the concepts of path integrals and techniques how to construct them are presented. Here we concentrate on a White Noise approach. Combining White Noise techniques with a generalized time-dependent Doss' formula Feynman integrands are constructed as white noise distributions beyond perturbation theory.

Keywords: Feynman Path Integrals, White Noise Analysis
PACS: 02.05.Fz, 03.65.Db

INTRODUCTION

The seminal idea of Feynman was an alternative approach to quantum mechanics by (path) integrals, see [1]. This method is often used in many branches of theoretical physics, e.g. quantum field theory and polymer physics. Unfortunately, Feynman's initial idea of averages over paths has a mathematical meaning only for the solutions of the heat equation. In this case one can present the solutions by a path integral, based on the Wiener measure. This is stated by the famous Feynman-Kac formula. There have been a lot of attempts to write down solutions of the Schrödinger equation as a Feynman (path) integral in a mathematical rigorous way. The methods used in this context (e.g. analytic continuation, limits of finite dimensional approximation and Fourier transform) are always more involved and less direct than in the euclidean – i.e. Feynman-Kac – case. Instead of giving a long list of publications with different approaches to Feynman integrals, we refer to [2] and the large number of references therein.

We choose a White Noise approach to show the existence of the Feynman integrand as a generalized function for a new class of potentials. White Noise is a framework which offers various generalizations of concepts known from finite-dimensional analysis, like differential operators and Fourier transform. Detailed information concerning these methods can be found in the monographs [3],[4],[5],[6] and the articles [7],[8],[9]. The idea to realize Feynman integrals within the White Noise framework was first mentioned in the work of Hida and Streit [10]. The concept of integral has a natural extension in the dual paring of generalized and test function and the construction of generalized function (the "Feynman integrand") follows closely Feynman's fundamental idea of a sum over histories. The Feynman integrand is an element of a suitable space of distributions chosen so as to accommodate various classes of admissible interaction potentials. For a construction see e.g. [5],[9],[11],[12],[13],[14],[15].

After recalling the idea of path integrals we give a brief introduction to the concepts of White Noise Analysis. Then by the picking up an idea of [11] and generalizing Doss'

CP1021, *Stochastic and Quantum Dynamics of Biomolecular Systems*,
Proceedings of the 5th Jagna International Workshop, edited by C. C. Bernido and M. V. Carpio-Bernido

formula to a time-dependent case, we illustrate how to construct Feynman integrands as white noise distributions beyond perturbation theory.

PATH INTEGRALS

As proposed by Feynman [16] quantum mechanical transition amplitudes may thought of as kind of averaging over fluctuating paths x, with oscillatory weight functions given in terms of the classical action

$$S(x) = \int_{t_0}^{t} L(x(\tau),\dot{x}(\tau),\tau)\,d\tau.$$

Typically the Lagrangian L is of the form

$$L(x,\dot{x},\cdot) = \frac{1}{2}m\dot{x}^2 - V(x,m\dot{x},\cdot),$$

for a particle of mass m moving in the force field of a potential V. Informally the Feynman path integral is then expressed by

$$K(t,y;t_0,y_0) = \mathrm{N}\int \exp\left(\frac{\mathrm{i}}{\hbar}\mathrm{S}(\mathrm{x})\right) \prod_{\mathrm{t}_0<\tau<\mathrm{t}} \mathrm{dx}(\tau), \quad \hbar = \frac{\mathrm{h}}{2\pi}.$$

Here h is Planck's constant and the integral is thought of as being over all paths x with $x(t_0) = y_0$ and $x(t) = y$. The quantum mechanical propagator $K(t,y;t_0,y_0)$ represents the quantum mechanical transposition amplitude for a particle to be found at position $y \in \mathbb{R}^d$ at time time t given that the particle was at position $y_0 \in \mathbb{R}^d$ at an earlier time t_0. From the mathematical point of view $K(t,y;t_0,y_0)$ is nothing else than the integral kernel of the corresponding evolution system $U(t,t_0)$. In detail

$$U(t,t_0)\Psi(y) = \int_{\mathbb{R}^d} K(t,y;t_0,u)\Psi(u)\,du, \quad U(t_0,t_0)\Psi = \Psi,$$

for a normalizable state Ψ, where the the evolution system $U(t,t_0)$ solves the Schrödinger equation

$$i\hbar\partial_t\Big(U(t,t_0)\Psi\Big) = H(t)\Big(U(t,t_0)\Psi\Big), \quad U(t_0,t_0)\Psi = \Psi, \tag{1}$$

corresponding to the Hamiltonian

$$H(t) = -\frac{\hbar^2}{2m}\Delta + V(q,\nabla,t),$$

for a potential V.

In the euclidean case, i.e., $t \mapsto -it$, the Feynman–Kac formula gives an explicit integral representation for the heat kernel

$$W(t,y;0,y_0) = \int \exp\left(-\int_0^t V(x(\tau))\,d\tau\right) \mathrm{N}\exp\left(-\frac{1}{2}\int_0^{\mathrm{t}} \dot{\mathrm{x}}(\tau)^2\,\mathrm{d}\tau\right) \prod_{0<\tau<\mathrm{t}} \mathrm{dx}(\tau)$$

$$= \int_\Omega \exp\left(-\int_0^t V(x(\tau))\,d\tau\right)\,dW_{y_0,y},$$

where $W_{y_0,y}$ is a Brownian Bridge from y_0 to y and $h = m = 1$. Moreover the semigroup $(T_t)_{t\geq 0}$, with generator

$$H = -\frac{1}{2}\Delta + V(q),$$

solves the heat equation with potential V

$$\partial_t\big(T_t\Psi\big) = -H\big(T_t\Psi\big), \quad T_0\Psi = \Psi,$$

and is given via the integral kernel (Green's function)

$$T_t\Psi(y) = \int_{\mathbb{R}^d} W(t,y;0,u)\Psi(u)\,du, \quad T_0\Psi = \Psi,$$

for $\psi \in D(H)$. For a large class of potentials, e.g. V bounded below, the Feynman–Kac formula can be realized in a mathematical rigorous way.

Here we use White Noise techniques to realize Feynman integrals. Our approach is related to the Feynman–Kac formula, because in both frameworks a Gaussian measure is taken as reference measure. In contrast, however, we work in physical time and have as integrand a generalized function instead of an integrable function.

WHITE NOISE ANALYSIS

The following section gives a short introduction to White Noise Analysis. For doing this we have to fix the following notations: The space of real-valued, square-integrable functions on $\mathbb{R}$ we denote by $L^2(\mathbb{R})$, $(\cdot,\cdot)$ and $\|\cdot\|_2 := \sqrt{(\cdot,\cdot)}$ denotes its usual scalar product and norm, respectively. Furthermore, the space of arbitrary often differentiable functions defined on $\mathbb{R}$ with compact support is denoted by $C_0^\infty(\mathbb{R})$.

We start by considering the Gel'fand triple

$$S(\mathbb{R}) \subset L^2(\mathbb{R}) \subset S'(\mathbb{R}),$$

where $S(\mathbb{R})$ is the space of real-valued, rapidly decreasing, smooth test functions on $\mathbb{R}$, equipped with its usual locally convex topology, see e.g. [17, Sec.V]. Its topological dual $S'(\mathbb{R})$ is the space of tempered distributions. The characteristic function

$$C_\mu(f) := \exp\left(-\tfrac{1}{2}\|f\|_2^2\right), \quad f \in S(\mathbb{R}),$$

leads us to the White Noise measure space $(S'(\mathbb{R}), \mathscr{B}, \mu)$ via Minlos' theorem, see e.g. [3],[4],[5]:

$$C_\mu(f) = \int_{S'} \exp\left(i\langle\omega, f\rangle\right)\,d\mu(\omega).$$

Here the dual pairing $\langle\cdot,\cdot\rangle$ of $S'(\mathbb{R})$ and $S(\mathbb{R})$ extends the inner product in $L^2(\mathbb{R})$, i.e., $\langle f,g\rangle = (f,g)$, $f \in L^2(\mathbb{R})$, $g \in S(\mathbb{R})$. We define the space $L^2(\mu)$ to be the Hilbert space $L^2(S'(\mathbb{R}),\mathscr{B},\mu)$ of complex-valued, square-integrable functions w.r.t. μ. In this context, in the sense of an $L^2(\mu)$-limit, a version of Wiener's Brownian motion is given by:

$$B(t,\omega) := \langle \omega, \mathbf{1}_{[0,t]}\rangle = \int_0^t \omega \, d\tau, \; t > 0, \; \omega \in S'(\mathbb{R}),$$

where $\mathbf{1}_A$ denotes the indicator function of $A \subset \mathbb{R}$. But for our considerations the space $L^2(\mu)$ is too small, so one introduces a space of test functionals in it to use its larger dual space (see [5] for a detailed construction), i.e., we get the Gel'fand triple:

$$(S) \subset L^2(\mu) \subset (S)'.$$

The elements of the space $(S)'$ are called Hida distributions (or generalized Brownian functionals). The so called S- or T-transforms provide a characterization of the spaces (S) and $(S)'$. These transforms are defined for $\Phi \in (S)'$ and $g \in S(\mathbb{R})$ as follows

$$(T\Phi)(g) := \langle\!\langle \Phi, \exp\left(i\langle\cdot,g\rangle\right)\rangle\!\rangle \quad \text{and} \quad (S\Phi)(g) := \langle\!\langle \Phi, :\exp\left(\langle\cdot,g\rangle\right):\rangle\!\rangle,$$

where $:\exp\left(\langle\cdot,g\rangle\right): := \exp\left(\langle\cdot,g\rangle - \frac{1}{2}\|g\|_2^2\right)$ is the Wick ordered exponential, and $\langle\!\langle\cdot,\cdot\rangle\!\rangle$ denotes the bilinear dual pairing between $(S)'$ and (S). Note that the functions $:\exp\left(\langle\cdot,g\rangle\right):$ and $\exp\left(i\langle\cdot,g\rangle\right)$ are in (S). If $\Phi \in L^2(\mathbb{R})$ then

$$(T\Phi)(g) = \int_{S'} \Phi(\omega)\exp\left(i\langle\cdot,g\rangle\right) d\mu(\omega).$$

For all $\Phi \in (S)'$ on has a generalized chaos decomposition, i.e.,

$$\Phi(\omega) = \sum_{n=0}^{\infty} \left\langle\!\!\left\langle :\omega^{\otimes n}:, \Phi^{(n)} \right\rangle\!\!\right\rangle, \quad \Phi^{(n)} \in (S)', \tag{2}$$

where $:\omega^{\otimes n}:$ denotes the n-th Wick power of $\omega \in S'(\mathbb{R})$, see e.g. [5], and $\Phi^{(n)}$ is called the kernel of its n-th chaos. Hence, by orthogonality of the Wick monomials one gets

$$S(\Phi)(g) = \sum_{n=0}^{\infty} \left\langle\!\!\left\langle \Phi^{(n)}, g^{\otimes n} \right\rangle\!\!\right\rangle, \quad g \in S(\mathbb{R}). \tag{3}$$

Furthermore the generalized expectation of a Hida distribution Φ is defined by $E_\mu(\Phi) := \langle\!\langle \Phi, 1\rangle\!\rangle$. It is possible to show for the S- and T- transforms that there exists an analytic extension to the complexified Schwartz space $S_{\mathbb{C}}(\mathbb{R})$. A relation of both is given by the following formula:

$$(T\Phi)(g) = C_\mu(g)(S\Phi)(ig) \quad \text{for all } g \in S_{\mathbb{C}}(\mathbb{R}). \tag{4}$$

Definition 0.1. *A function $F: S(\mathbb{R}) \to \mathbb{C}$ is called U-functional if it satisfies the following conditions:*

(i) *F is "ray-analytic", i.e., for all $g, f \in S(\mathbb{R})$ the mapping*

$$\mathbb{R} \ni x \mapsto F(g + xf) \in \mathbb{C}$$

has an analytic continuation to $\mathbb{C}$ as an entire function.

(ii) *F is uniformly bounded of order 2, i.e., there exist some constants $0 < K, D < \infty$ and a continuous norm $\|\cdot\|$ on $S(\mathbb{R})$ such that*

$$|F(zg)| \leq K \exp(D|z|^2 \|g\|^2),$$

for all $z \in \mathbb{C}$, $g \in S(\mathbb{R})$.

Theorem 0.2. *The following statements are equivalent:*

(i) *$F: S(\mathbb{R}) \to \mathbb{C}$ is a U-functional.*

(ii) *F is the S-transform of a unique Hida distribution $\Phi \in (S)'$.*

(iii) *F is the T-transform of a unique Hida distribution $\Phi \in (S)'$.*

For the proof and more detailed information see e.g. [5].

FEYNMAN INTEGRAND AS A WHITE NOISE DISTRIBUTION

In this section we first give a general strategy to construct Feynman integrands in White Noise Analysis. After that we consider non-pertubative accessible potentials with help of a complex scaling method. Here we restricted ourself to the one dimensional case $d = 1$.

Strategy

We are interested in constructing the Feynman integrand corresponding to a potential V as a Hida distribution I_V such that $E_\mu(I_V) = K_V(t, y; t_0, y_0)$, $y, y_0 \in \mathbb{R}$, $t > t_0$. A general Gaussian ansatz is of the form

$$I_V = \mathrm{Nexp}\left(\frac{\mathrm{i}}{2}\int_{t_0}^{t} \dot{\mathrm{x}}(\tau)^2 \, \mathrm{d}\tau + \frac{1}{2}\int_{t_0}^{t} \dot{\mathrm{B}}_\tau^2 \, \mathrm{d}\tau\right) \cdot \mathrm{Nexp}\left(-\mathrm{i}\int_{t_0}^{t} \mathrm{V}(\mathrm{x}(\tau), \dot{\mathrm{x}}(\tau), \tau) \, \mathrm{d}\tau\right) \cdot \delta(\mathrm{x}(\mathrm{t}) - \mathrm{y}), \quad (5)$$

where Nexp is a normalized exponential, $\frac{i}{2}\int_{t_0}^{t} \dot{x}(\tau)^2 \, d\tau$ corresponds to the kinetic energy, $\frac{i}{2}\int_{t_0}^{t} \dot{x}(\tau)^2 \, d\tau$ is used to compensate the Gaussian fall off between t_0 and t and $-i\int_{t_0}^{t} V(x(\tau), \dot{x}(\tau), \tau) \, d\tau$ takes the potential energy into account. In the White Noise framework (5) can be realized as

$$I_V = \mathrm{Nexp}\left(\frac{\mathrm{i}}{2}\int_{\mathrm{t}_0}^{\mathrm{t}} \omega^2 \, \mathrm{d}\tau + \frac{1}{2}\int_{\mathrm{t}_0}^{\mathrm{t}} \langle \omega, \mathbf{1}_{[\mathrm{t}_0,\mathrm{t}]}\rangle^2 \, \mathrm{d}\tau\right)$$

$$\cdot \mathrm{Nexp}\left(-\mathrm{i}\int_{\mathrm{t}_0}^{\mathrm{t}} \mathrm{V}(\langle \omega, \mathbf{1}_{[\mathrm{t}_0,\mathrm{t}]}\rangle, \omega, \tau) \, \mathrm{d}\tau\right) \cdot \delta(\langle \omega, \mathbf{1}_{[\mathrm{t}_0,\mathrm{t}]}\rangle + \mathrm{y}_0 - \mathrm{y}). \quad (6)$$

In the case of free motion, i.e. $V = 0$, one easily shows that $I_0 \in (S)$ and one can explicitly calculate that

$$E_\mu(I_0) = \left(\frac{1}{i2\pi(t-t_0)}\right)^{1/2} \exp\left(\frac{i(y-y_0)^2}{2(t-t_0)}\right) = K_0(t,y;t_0,y_0),$$

see [5]. This has been extended to general quadratic actions, see [11] and [18], where one still has an explicit formula for $K_V(t,y;t_0,y_0)$. The T-transform of the corresponding Feynman integrands were a starting point for treating a huge class of potentials V by perturbation techniques, see [5],[9],[12],[13],[14],[15]. In all cases the corresponding Feynman integrand exists as a Hida distribution or is in the larger space of Kondratiev distributions. In the following theorem we show how to realize the Feynman Integrand as a Hida distribution beyond perturbation theory. Assuming that $I_V \in (S)'$ exists, it possesses a generalized chaos decomposition, see (2). By (6) it is natural to assume that

$$\mathrm{supp}(I_V^{(n)}) \subset [t_0,t]^n \quad \text{for all } n \in \mathbb{N}, \tag{7}$$

where $I_V^{(n)}$ is the kernel of the n-th chaos of I_V.

Theorem 0.3. *Let V be a potential on $\mathbb{R}$ such that I_V as in (6) exists as a Hida distribution. Moreover we assume that the kernels of its generalized chaos decomposition have the support property as in (7). Then:*

$$\exp\left(-iyg(t) + iy_0 g(t_0) + \frac{1}{2}\int_{\mathbb{R}\setminus[t_0,t]} g^2 \, d\tau\right) TI_V(g)$$
$$= E_\mu(I_{V+\dot{g}(t)q}) = K_{V+\dot{g}(t)q}(t,y;t_0,y_0).$$

Proof. Let $y,y_0 \in \mathbb{R}$ and $t,t_0 \in \mathbb{R}$, $t > t_0$, then by (4) together with (3) for $g \in S(\mathbb{R})$ one has that

$$\begin{aligned}
TI_V(g) &= \langle\langle I_V, \exp(i\langle \cdot, g\rangle)\rangle\rangle \\
&= \langle\langle I_V, :\exp(i\langle \cdot, g\rangle):\rangle\rangle \exp\left(-\frac{1}{2}\int_{\mathbb{R}} g^2 \, d\tau\right) \\
&= \sum_{n=0}^{\infty} \left\langle\left\langle I_V^{(n)}, g^{\otimes n}\right\rangle\right\rangle \exp\left(-\frac{1}{2}\int_{\mathbb{R}} g^2 \, d\tau\right) \\
&= \sum_{n=0}^{\infty} \left\langle\left\langle I_V^{(n)}, (g\mathbf{1}_{[t_0,t]})^{\otimes n}\right\rangle\right\rangle \exp\left(-\frac{1}{2}\int_{\mathbb{R}} g^2 \, d\tau\right) \\
&= \langle\langle I_V, :\exp(i\langle \cdot, g\mathbf{1}_{[t_0,t]}\rangle):\rangle\rangle \exp\left(-\frac{1}{2}\int_{\mathbb{R}} g^2 \, d\tau\right)
\end{aligned}$$

$$= E_\mu \left(I_V \cdot \exp\left(i \int_{t_0}^{t} \omega g \, d\tau \right) \right) \exp\left(-\frac{1}{2} \int_{\mathbb{R}\setminus[t_0,t]} g^2 \, d\tau \right),$$

because $\mathrm{supp}(\mathrm{I}_{\mathrm{V}}^{(\mathrm{n})}) \in [\mathrm{t}_0, \mathrm{t}]^{\mathrm{n}}$ for all $n \in \mathbb{N}$. Furthermore, an integration by parts yields

$$TI_V(g) = E_\mu \left(I_V \cdot \exp\left(-i \int_{t_0}^{t} x(\tau)\dot{g}(\tau) \, d\tau \right) \right) \cdot \exp\left(+iyg(t) - iy_0 g(t_0) - \frac{1}{2} \int_{\mathbb{R}\setminus[t_0,t]} g^2 \, d\tau \right).$$

Hence we get that

$$\exp\left(-iyg(t) + iy_0 g(t_0) + \frac{1}{2} \int_{\mathbb{R}\setminus[t_0,t]} g^2 \, d\tau \right) TI_V(g) = E_\mu(I_{V+\dot{g}(t)q}) = K_{V+\dot{g}(t)q}(t,y;t_0,y_0).$$

□

Beyond Perturbation Theory

Let $T > 0$, $\mathscr{O} \subset \mathbb{R}$ such that $\mathbb{R} \setminus \mathscr{O}$ is of measure zero and $\mathscr{D} = \{y + \sqrt{i}u,\ y \in \mathscr{O},\ u \in \mathbb{R}\}$. Moreover let $V_0 : \mathscr{D} \to \mathbb{C}$ be an analytic potential and $\Psi : \mathbb{C} \to \mathbb{C}$ an analytic function. Then for $y \in \mathscr{O}$ and $t \in [0,T]$ it is show in [19] that under suitable conditions the operator

$$T_t\Psi(y) := \int_\Omega \exp\left(-i \int_0^t V_0(y + \sqrt{i}B_\tau) \, d\tau \right) \Psi(y + \sqrt{i}B_t) \, dW, \qquad t \in [0,T],$$

where W is the law of a Brownian motion starting at 0, is the unique solution to the Schrödinger equation

$$i\partial_t \big(T_t\Psi\big)(y) = H\big(T_t\Psi\big)(y), \quad T_0\Psi = \Psi.$$

Here the Hamiltonian to a potential V_0 is of the form

$$H = -\frac{1}{2}\Delta + V_0(q).$$

Motivated by Theorem 0.3 we are interested in a generalization to time-dependent potentials of the form

$$V_{\dot{g}}(z,t) = V_0(z) + \dot{g}(t)z, \quad g \in S(\mathbb{R}), \quad z \in \mathscr{D}.$$

So for $0 < t_0 < t < T$, $z \in \mathscr{D}$ and $g \in S(\mathbb{R})$ we consider the time-dependent operator

$$U_{\dot{g}}(t,t_0)\Psi(z) := \int_\Omega \exp\left(-i \int_0^{t-t_0} V_{\dot{g}}(z + \sqrt{i}B_\tau, t-\tau) \, d\tau \right) \Psi(z + \sqrt{i}B_{t-t_0}) \, dW, \qquad (8)$$

where W is again the law of a Brownian motion starting at 0.

Theorem 0.4. *Let V_0 be a potential of the form*

$$V_0(z) = (-1)^{n+1} a_{4n+2} z^{4n+2} + \sum_{j=1}^{4n+1} a_j z^j, \quad z \in \mathbb{C},\ n \in \mathbb{N},$$

for $a_{4n+2} > 0$, $a_j \in \mathbb{C}, j \in \{0,..,4n+1\}$, or

$$V_0(z) = \frac{1}{|z|^m}, \quad z \neq 0,\ m \in \mathbb{N},$$

and Ψ a Hermite function. Then:

(i) *The generalization of Doss' formula (8) to the corresponding time-dependent potentials $V_{\dot{g}}$ is well-defined for all $z \in \mathscr{D}$.*

(ii) *$U_{\dot{g}}(t,t_0)\Psi(y)$ solves the Schrödinger equation corresponding to the Hamiltonian $H(t) = -\frac{1}{2}\Delta + V_0 + \dot{g}(t)q$ for $y \in \mathscr{O}$.*

(iii) *$g \mapsto U_{\dot{g}}(t,t_0)\Psi(y)$ is a U-functional for $y \in \mathscr{O}$. Hence by Theorem 0.2 there exists the Feynman integrand corresponding to such V_0 and Ψ as a Hida distribution in the sense of Theorem 0.3.*

For the proof and a detailed construction see [20].

ACKNOWLEDGMENTS

Dear Ludwig Streit, dear Frederik W. Wiegel, we wish you both all the best to your seventieth birthday. Martin Grothaus would like to thank M. Victoria Carpio-Bernido and Christopher C. Bernido for their warm hospitality during the 5th Jagna International Workshop in January 2008.

REFERENCES

1. R. Feynman, and A. Hibbs, *Quantum Mechanics and Path Integrals*, McGraw-Hill, London, New York, 1965.
2. S. Albeverio, R. Høegh-Krohn, and S. Mazzucchi, *Mathematical Theory of Feynman Path Integrals: An Introduction*, vol. 523 of *Lecture Notes in Mathematics*, Springer Verlag, Berlin, Heidelberg, New York, 2008.
3. T. Hida, *Brownian Motion*, Springer Verlag, Berlin, Heidelberg, New York, 1980.
4. Y. Berezansky, and Y. Kondratiev, *Spectral Methods in Infinite-Dimensional Analysis*, Kluwer Academic Publishers, Dordrecht, 1995, originally in Russian, Naukova Dumka, Kiev, 1988.
5. T. Hida, H.-H. Kuo, J. Potthoff, and L. Streit, *White Noise. An infinite dimensional calculus*, Kluwer Academic Publisher, Dordrecht, Boston, London, 1993.
6. H.-H. Kuo, *White Noise Distribution Theory*, CRC Press, Boca Raton, New York, London, Tokyo, 1996.
7. J. Potthoff, "Introduction to white noise analysis," in *Control theory, stochastic analysis and applications (Hangzhou, 1991)*, World Sci. Publ., River Edge, NJ, 1991, pp. 39–58.
8. Y. Kondratiev, P. Leukert, J. Potthoff, L. Streit, and W. Westerkamp, *J. Funct. Anal.* **141**, 301–318 (1996).

9. J. L. Silva, and L. Streit, "Feynman integrals and white noise analysis," in *Stochastic analysis and mathematical physics (SAMP/ANESTOC 2002)*, World Sci. Publ., River Edge, NJ, 2004, pp. 285–303.
10. T. Hida, and L. Streit, *Stoch. Proc. Appl.* **16**, 55–69 (1983).
11. M. de Faria, J. Potthoff, and L. Streit, *J. Math. Phys.* **32**, 2123–2127 (1991).
12. D. Khandekar, and L. Streit, *Ann. Physik* **1**, 46–55 (1992).
13. M. Grothaus, D. Khandekar, J. Silva, and L. Streit, *J. Math. Phys.* **38**, 3278–3299 (1997).
14. T. Kuna, L. Streit, and W. Westerkamp, *J. Math. Phys.* **39**, 4476–4491 (1998).
15. M. de Faria, M. J. Oliveira, and L. Streit, *J. Math. Phys.* **46**, 063505, 14 (2005), ISSN 0022-2488.
16. R. P. Feynman, *Rev. Modern Physics* **20**, 367–387 (1948), ISSN 0034-6861.
17. M. Reed, and B. Simon, *Methods of modern mathematical physics*, vol. I, Academic Press, New York, London, 1975.
18. M. Grothaus, and L. Streit, *Rep. Math. Phys.* **44**, 381–405 (1999).
19. H. Doss, *Communications in Mathematical Physics* **73**, 247–264 (1980).
20. M. Grothaus, L. Streit, and A. Vogel (2008), submitted to the festschrift in honor of Jean-Michel Bismut's sixtieth birthday.

Intersection local times of fractional Brownian motions with $H \in (0,1)$ as generalized white noise functionals

Custódia Drumond*, Maria João Oliveira† and José Luís da Silva*

*DME and CCM, University of Madeira, P 9000-390 Funchal, Portugal
†Universidade Aberta, P 1269-001 Lisbon, Portugal; CMAF, University of Lisbon, P 1649-003 Lisbon, Portugal

Abstract. In $\mathbb{R}^d$, for any dimension $d \geq 1$, expansions of self-intersection local times of fractional Brownian motions with arbitrary Hurst coefficients in $(0,1)$ are presented. The expansions are in terms of Wick powers of white noises (corresponding to multiple Wiener integrals), being well-defined in the sense of generalized white noise functionals.

Keywords: Fractional Brownian motion; Fractional white noise analysis; Local time
PACS: 02.30.Sa, 02.30.Mv, 05.40.Jc

Dedicated to Professor Ludwig Streit on the occasion of his 70th Birthday

INTRODUCTION

Intersection local times of fractional Brownian motion (fBm) has been studied by many authors, see e.g. the works done by Gradinaru et al. [5], Nualart et al. [9], [8], Rosen [21], and the references therein.

We may consider intersections of simple paths of a fBm with themselves or with other independent fractional Brownian motion as in [16]. Within the white noise analysis framework, we concentrate here in the simple self-intersection problem using a similar approach to Faria et al. [4], [3] to study the case $H = 1/2$ (the classical Brownian motion). This approach has the advantage that the underlying probability space does not depend on the Hurst coefficient under consideration. As a consequence, we analyze the self-intersections of a fBm, with Hurst coefficient being any possible value in $(0,1)$. As one may expect, independently of the Hurst coefficient under consideration, self-intersections become scarce as the dimension d increases.

An informal but suggestive definition of self-intersection local time of a generic d-dimensional fBm B is in terms of an integral over Donsker's δ-function

$$L = \int d^2t\, \delta(B(t_2) - B(t_1)),$$

intended to measure the amount of "time" the process spends intersecting itself.

CP1021, *Stochastic and Quantum Dynamics of Biomolecular Systems, Proceedings of the 5th Jagna International Workshop,* edited by C. C. Bernido and M. V. Carpio-Bernido

A rigorous definition, such as, e.g., through a sequence of Gaussians approximating the δ-function,

$$(2\pi\varepsilon)^{-d/2}\exp\left(-\frac{|x|^2}{2\varepsilon}\right), \quad \varepsilon > 0,$$

will make L increasingly singular, and various "renormalizations" have to be done as the dimension d increases. For $d > 1$ and $1/d \leq H < 3/(2+d)$, the expectation diverges in the limit and must be subtracted. The L^2-properties related to this situation have been analyzed in [8]. For other H values further kernel terms must be subtracted (Theorem 7 below).

In this work we are particularly interested in the chaos decomposition of L. We expand L in terms of Wick powers [7] of white noise, an expansion which corresponds to that in terms of multiple Wiener integrals when one considers the Wiener process as the fundamental random variable. This allows us to derive the kernels for L. Due to the local structure of the Wick powers, the kernel functions are relatively simple and exhibit clearly the dimension dependence singularities of L (Proposition 8). For comparison, we also calculate the regularized kernel functions corresponding to the Gaussian δ-sequence mentioned above.

The paper is organized as follows. In Section 2 we review the necessary background of white noise analysis and in Section 3 we present the main results of this work and their proofs. We shall mention that after the conclusion of our work we found the recent paper [20] with related results.

GAUSSIAN WHITE NOISE CALCULUS

In this section we briefly recall the concepts and results of white noise analysis used throughout this work (for a detailed explanation see e.g. [2], [6], [7], [10], [12], [13], [14], [15], [17]).

Fractional Brownian motion

The starting point of white noise analysis is the real Gelfand triple

$$S_d(\mathbb{R}) \subset L^2_d(\mathbb{R}) \subset S'_d(\mathbb{R}),$$

where $L^2_d(\mathbb{R}) := L^2(\mathbb{R},\mathbb{R}^d)$, $d \geq 1$, is the real Hilbert space of all vector valued square integrable functions with respect to the Lebesgue measure on $\mathbb{R}$, $S_d(\mathbb{R})$ and $S'_d(\mathbb{R})$ are the Schwartz spaces of the vector valued test functions and tempered distributions, respectively. We shall denote the $L^2_d(\mathbb{R})$-norm by $|\cdot|$ and the dual pairing between $S'_d(\mathbb{R})$ and $S_d(\mathbb{R})$ by $\langle\cdot,\cdot\rangle$, which is defined as the bilinear extension of the inner product on $L^2_d(\mathbb{R})$, i.e.,

$$\langle \mathbf{g},\mathbf{f}\rangle = \sum_{i=1}^{d}\int_{\mathbb{R}} dx\, g_i(x) f_i(x),$$

for all $\mathbf{g} = (g_1, ..., g_d) \in L^2_d(\mathbb{R})$ and all $\mathbf{f} = (f_1, ..., f_d) \in S_d(\mathbb{R})$. By the Minlos theorem, there is a unique probability measure μ on the σ-algebra $\mathscr{B}$ generated by the cylinder sets on $S'_d(\mathbb{R})$ with characteristic function given by

$$C(\mathbf{f}) := \int_{S'_d(\mathbb{R})} d\mu(\omega) e^{i\langle \omega, \mathbf{f}\rangle} = e^{-\frac{1}{2}|\mathbf{f}|^2}, \quad \mathbf{f} \in S_d(\mathbb{R}).$$

In this way we have defined the white noise measure space $(S'_d(\mathbb{R}), \mathscr{B}, \mu)$. Within this formalism, a version of the d-dimensional Wiener Brownian motion is given by

$$\mathbf{B}(t) := \left(\langle \omega_1, 1\!\!1_{[0,t]}\rangle, ..., \langle \omega_d, 1\!\!1_{[0,t]}\rangle\right), \quad (\omega_1, ..., \omega_d) \in S'_d(\mathbb{R})$$

where $1\!\!1_A$ denotes the indicator function of a set A. For an arbitrary Hurst parameter $0 < H < 1, H \neq \frac{1}{2}$, a version of a d-dimensional fractional Brownian motion is given by

$$\mathbf{B}_H(t) := \left(\langle \omega_1, M_H 1\!\!1_{[0,t]}\rangle, ..., \langle \omega_d, M_H 1\!\!1_{[0,t]}\rangle\right), \quad (\omega_1, ..., \omega_d) \in S'_d(\mathbb{R}),$$

where, for a generic real valued function f, and for $\frac{1}{2} < H < 1$,

$$(M_H f)(x) := \frac{K_H}{\Gamma\left(H - \frac{1}{2}\right)} \int_x^\infty dy\, f(y)(y - x)^{H - \frac{3}{2}}, \tag{1}$$

provided the integral exists for all $x \in \mathbb{R}$, while for $0 < H < \frac{1}{2}$,

$$(M_H f)(x) := \frac{(\frac{1}{2} - H)K_H}{\Gamma\left(H + \frac{1}{2}\right)} \lim_{\varepsilon \to 0^+} \int_\varepsilon^\infty dy \frac{f(x) - f(x + y)}{y^{\frac{3}{2} - H}}, \tag{2}$$

provided the limit exists for almost all $x \in \mathbb{R}$ (for more details see e.g. [1] and [19] and the references therein). Independently of the case under consideration, the normalizing constant K_H is given by

$$K_H = \Gamma\left(H + \frac{1}{2}\right) \left(\frac{1}{2H} + \int_0^\infty ds \left((1 + s)^{H - \frac{1}{2}} - s^{H - \frac{1}{2}}\right)\right)^{-\frac{1}{2}}.$$

Apart from the operators (1), (2), we shall also consider the operator defined for $\frac{1}{2} < H < 1$ by

$$(M_H^+ f)(x) := \frac{K_H}{\Gamma\left(H - \frac{1}{2}\right)} \int_{-\infty}^x dy\, f(y)(x - y)^{H - \frac{3}{2}}, \tag{3}$$

provided the integral exists for all $x \in \mathbb{R}$, and the operator defined for $0 < H < \frac{1}{2}$ by

$$(M_H^+ f)(x) := \frac{(\frac{1}{2} - H)K_H}{\Gamma\left(H + \frac{1}{2}\right)} \lim_{\varepsilon \to 0^+} \int_\varepsilon^\infty dy \frac{f(x) - f(x - y)}{y^{\frac{3}{2} - H}}, \tag{4}$$

provided the limit exists for almost all $x \in \mathbb{R}$.

There are several examples of functions f for which $M_H f$ and $M_H^+ f$ exist for any $H \in (0,1)$. For instance, $f = 1\!\!1_{[0,t]}$, $t > 0$, or $f \in S_1(\mathbb{R})$. For functions f_1, f_2 being either one of these two types it is easy to prove the following equality

$$\int_{\mathbb{R}} dx\, f_1(x)(M_H f_2)(x) = \int_{\mathbb{R}} dx\, (M_H^+ f_1)(x) f_2(x),$$

showing that M_H and M_H^+ are dual operators. For more details and proofs see e.g. [1] and the references therein.

Hida distributions and characterization results

Let us now consider the complex Hilbert space $L^2(S'_d(\mathbb{R}), \mathscr{B}, \mu)$. This space is canonically isomorphic to the symmetric Fock space of symmetric square integrable functions,

$$L^2(S'_d(\mathbb{R}), \mathscr{B}, \mu) \simeq \left(\bigoplus_{k=0}^{\infty} \mathrm{Sym}\, L^2(\mathbb{R}^k, k! d^k x) \right)^{\otimes d},$$

leading to the chaos expansion of the elements in $L^2(S'_d(\mathbb{R}), \mathscr{B}, \mu)$,

$$F(\omega_1, \ldots, \omega_d) = \sum_{(n_1, \ldots, n_d) \in \mathbb{N}^d} \langle : \omega_1^{\otimes n_1} : \otimes \cdots \otimes : \omega_d^{\otimes n_d} :, \mathbf{f}_{(n_1, \ldots, n_d)} \rangle, \tag{5}$$

with kernel functions $\mathbf{f}_{(n_1, \ldots, n_d)}$ in the Fock space. For simplicity, in the sequel we shall use the notation

$$\mathbf{n} = (n_1, \cdots, n_d) \in \mathbb{N}^d, \quad n = \sum_{i=1}^{d} n_i, \quad \mathbf{n}! = \prod_{i=1}^{d} n_i!,$$

which reduces expansion (5) to

$$F(\omega) = \sum_{\mathbf{n} \in \mathbb{N}^d} \langle : \omega^{\otimes \mathbf{n}} :, \mathbf{f}_{\mathbf{n}} \rangle, \quad \omega \in S'_d(\mathbb{R}).$$

To proceed further we have to consider a Gelfand triple around the space $L^2(S'_d(\mathbb{R}), \mathscr{B}, \mu)$. We will use the space $(S)^*$ of Hida distributions (or generalized Brownian functionals) and the corresponding Gelfand triple $(S) \subset L^2(S'_d(\mathbb{R}), \mathscr{B}, \mu) \subset (S)^*$. Here (S) is the space of white noise test functions such that its dual space (with respect to $L^2(S'_d(\mathbb{R}), \mathscr{B}, \mu)$) is the space $(S)^*$. Instead of reproducing the explicit construction of $(S)^*$ (see e.g. [7]), in Theorem 2 below we define this space by its S-transform. We recall that given a $\mathbf{f} \in S_d(\mathbb{R})$, and the Wick exponential

$$: \exp(\langle \omega, \mathbf{f} \rangle) : := \sum_{\mathbf{n} \in \mathbb{N}^d} \frac{1}{\mathbf{n}!} \langle : \omega^{\otimes \mathbf{n}} :, \mathbf{f}^{\otimes \mathbf{n}} \rangle = C(\mathbf{f}) e^{\langle \omega, \mathbf{f} \rangle},$$

we define the S-transform of a $\Phi \in (S)^*$ by

$$S\Phi(\mathbf{f}) := \langle\!\langle \Phi, : \exp(\langle \cdot, \mathbf{f} \rangle) : \rangle\!\rangle, \quad \forall \mathbf{f} \in S_d(\mathbb{R}). \tag{6}$$

Here $\langle\langle\cdot,\cdot\rangle\rangle$ denotes the dual pairing between $(S)^*$ and (S) which is defined as the bilinear extension of the sesquilinear inner product on $L^2(S'_d(\mathbb{R}),\mathscr{B},\mu)$. We observe that the multilinear expansion of (6),

$$S\Phi(\mathbf{f}) := \sum_{\mathbf{n}} \langle F_{\mathbf{n}}, \mathbf{f}^{\otimes \mathbf{n}} \rangle,$$

extends the chaos expansion to $\Phi \in (S)^*$ with distribution valued kernels $F_{\mathbf{n}}$ such that

$$\langle\langle \Phi, \varphi \rangle\rangle = \sum_{\mathbf{n}} \mathbf{n}! \langle F_{\mathbf{n}}, \varphi_{\mathbf{n}} \rangle,$$

for every generalized test function $\varphi \in (S)$ with kernel functions $\varphi_{\mathbf{n}}$.

In order to characterize the space $(S)^*$ through its S-transform we need the following definition.

Definition 1 *A function $F : S_d(\mathbb{R}) \to \mathbb{C}$ is called a U-functional whenever*
1. for every $\mathbf{f}_1, \mathbf{f}_2 \in S_d(\mathbb{R})$ the mapping $\mathbb{R} \ni \lambda \longmapsto F(\lambda \mathbf{f}_1 + \mathbf{f}_2)$ has an entire extension to $\lambda \in \mathbb{C}$,
2. there are constants $K_1, K_2 > 0$ such that

$$|F(z\mathbf{f})| \leq K_1 \exp\left(K_2 |z|^2 \|\mathbf{f}\|^2\right), \quad \forall z \in \mathbb{C}, \mathbf{f} \in S_d(\mathbb{R})$$

for some continuous norm $\|\cdot\|$ on $S_d(\mathbb{R})$.

We are now ready to state the aforementioned characterization result.

Theorem 2 ([11], [18]) *The S-transform defines a bijection between the space $(S)^*$ and the space of U-functionals.*

As a consequence of Theorem 2 one may derive the next two statements. The first one concerns the convergence of sequences of Hida distributions and the second one the Bochner integration of families of the same type of distributions (for more details and proofs see e.g. [7], [11], [18]).

Corollary 3 *Let $(\Phi_n)_{n \in \mathbb{N}}$ be a sequence in $(S)^*$ such that*

(i) For all $\mathbf{f} \in S_d(\mathbb{R})$, $((S\Phi_n)(\mathbf{f}))_{n \in \mathbb{N}}$ is a Cauchy sequence in $\mathbb{C}$,
(ii) There are $K_1, K_2 > 0$ such that for some continuous norm $\|\cdot\|$ on $S_d(\mathbb{R})$ one has

$$|(S\Phi_n)(z\mathbf{f})| \leq K_1 e^{K_2 |z|^2 \|\mathbf{f}\|^2}, \quad \forall z \in \mathbb{C}, \mathbf{f} \in S_d(\mathbb{R}), n \in \mathbb{N}.$$

Then $(\Phi_n)_{n \in \mathbb{N}}$ converges strongly in $(S)^*$ to an unique Hida distribution.

Corollary 4 *Let $(\Omega, \mathscr{B}, m)$ be a measure space and $\lambda \mapsto \Phi_\lambda$ be a mapping from Ω to $(S)^*$. We assume that the S-transform of Φ_λ fulfills the following two properties:*

(i) The mapping $\lambda \mapsto (S\Phi_\lambda)(\mathbf{f})$ is measurable for every $\mathbf{f} \in S_d(\mathbb{R})$,

(ii) The $S\Phi_\lambda$ obeys a U-estimate

$$|(S\Phi_\lambda)(z\mathbf{f})| \leq C_1(\lambda)e^{C_2(\lambda)|z|^2\|\mathbf{f}\|^2}, \quad z \in \mathbb{C}, \mathbf{f} \in S_d(\mathbb{R}),$$

for some continuous norm $\|\cdot\|$ on $S_d(\mathbb{R})$ and for $C_1 \in L^1(\Omega, m)$, $C_2 \in L^\infty(\Omega, m)$.

Then

$$\int_\Omega dm(\lambda)\Phi_\lambda \in (S)^*,$$

and

$$S\left(\int_\Omega dm(\lambda)\Phi_\lambda\right)(\mathbf{f}) = \int_\Omega dm(\lambda)S\Phi_\lambda(\mathbf{f}).$$

INTERSECTION LOCAL TIMES

Proposition 5 *For $t \neq s$ the Bochner integral*

$$\delta(\mathbf{B}_H(t) - \mathbf{B}_H(s)) := \left(\frac{1}{2\pi}\right)^d \int_{\mathbb{R}^d} d\lambda\, e^{i\lambda(\mathbf{B}_H(t)-\mathbf{B}_H(s))} \tag{7}$$

is a Hida distribution with S-transform given at each $\mathbf{f} \in S_d(\mathbb{R})$ by

$$S\delta(\mathbf{B}_H(t) - \mathbf{B}_H(s))(\mathbf{f}) = \left(\frac{1}{\sqrt{2\pi}|t-s|^H}\right)^d e^{-\frac{1}{2|t-s|^{2H}}\left|\int_{\mathbb{R}} dx\,\mathbf{f}(x)(M_H\mathbf{1}_{[s\wedge t, s\vee t]})(x)\right|^2}. \tag{8}$$

Proof. The proof of this result follows from the application of Corollary 4 to the S-transform of the integrand function in (7) with respect to the Lebesgue measure on $\mathbb{R}^d$.

Since

$$Se^{i\lambda(\mathbf{B}_H(t)-\mathbf{B}_H(s))}(\mathbf{f}) = e^{-\frac{|\lambda|^2}{2}(t-s)^{2H}} e^{i\lambda\int_{\mathbb{R}} dx\,\mathbf{f}(x)(M_H\mathbf{1}_{[s,t]})(x)}, \tag{9}$$

for $t > s$, and

$$Se^{i\lambda(\mathbf{B}_H(t)-\mathbf{B}_H(s))}(\mathbf{f}) = e^{-\frac{|\lambda|^2}{2}(s-t)^{2H}} e^{-i\lambda\int_{\mathbb{R}} dx\,\mathbf{f}(x)(M_H\mathbf{1}_{[t,s]})(x)}, \tag{10}$$

for $t < s$ cf. e.g. [7], clearly in both situations the measurability condition is fulfilled.

Independently of the particular case under consideration, we observe that for all $z \in \mathbb{C}$ and all $\mathbf{f} \in S_d(\mathbb{R})$ we find

$$\begin{aligned}\left|Se^{i\lambda(\mathbf{B}_H(t)-\mathbf{B}_H(s))}(z\mathbf{f})\right| &= e^{-\frac{|\lambda|^2}{4}|t-s|^{2H}}\left|e^{-\frac{|\lambda|^2}{4}|t-s|^{2H}\pm iz\lambda\int_{\mathbb{R}} dx\,\mathbf{f}(x)(M_H\mathbf{1}_{[s\wedge t,s\vee t]})(x)}\right| \\ &\leq e^{-\frac{|\lambda|^2}{4}|t-s|^{2H}} e^{-\frac{|\lambda|^2}{4}|t-s|^{2H}+|z||\lambda|\left|\int_{\mathbb{R}} dx\,\mathbf{f}(x)(M_H\mathbf{1}_{[s\wedge t,s\vee t]})(x)\right|},\end{aligned}$$

where the latter exponential is bounded by

$$e^{\frac{|z|^2}{|t-s|^{2H}}\left|\int_{\mathbb{R}} dx\,\mathbf{f}(x)(M_H\mathbf{1}_{[s\wedge t,s\vee t]})(x)\right|^2}.$$

As a result,

$$\left|Se^{i\lambda(\mathbf{B}_H(t)-\mathbf{B}_H(s))}(z\mathbf{f})\right| \leq e^{-\frac{|\lambda|^2}{4}|t-s|^{2H}} e^{\frac{|z|^2}{|t-s|^{2H}}\left|\int_{\mathbb{R}} dx \mathbf{f}(x)(M_H \mathbf{1}_{[s\wedge t, s\vee t]})(x)\right|^2},$$

where, as a function of λ, the first factor is integrable on $\mathbb{R}^d$ and the second factor is constant.

An application of the result mentioned above completes the proof. In particular, it yields (8) by integrating (9), (10) over λ. ■

In order to proceed further the next result shows to be very useful. It improves the estimate obtained in [1, Theorem 2.3] towards the characterization results stated in Corollaries 3 and 4.

Lemma 6 *Let $H \in (0,1)$ and $f \in S_1(\mathbb{R})$ be given. There is a non-negative constant C_H independent of f such that*

$$\left|\int_{\mathbb{R}} dx\, f(x)(M_H \mathbb{1}_{[s,t]})(x)\right| \leq C_H|t-s|\left(\sup_{x\in\mathbb{R}}|f(x)| + \sup_{x\in\mathbb{R}}|f'(x)| + |f|\right)$$

for all $s < t$.

Proof. Since M_H and M_H^+ are dual operators,

$$\int_{\mathbb{R}} dx\, f(x)(M_H \mathbb{1}_{[s,t]})(x) = \int_s^t dx\,(M_H^+ f)(x),$$

and thus

$$\left|\int_{\mathbb{R}} dx\, f(x)(M_H \mathbb{1}_{[s,t]})(x)\right| \leq |t-s| \sup_{x\in\mathbb{R}} |(M_H^+ f)(x)|.$$

It has been shown in [1, Proof of Theorem 2.3] that for $0 < H < \frac{1}{2}$,

$$\sup_{x\in\mathbb{R}} |(M_H^+ f)(x)| \leq C'_H \left(\frac{2}{\frac{1}{2}-H}\sup_{x\in\mathbb{R}}|f(x)| + \frac{1}{H+\frac{1}{2}}\sup_{x\in\mathbb{R}}|f'(x)|\right),$$

for some constant $C'_H > 0$ independent of f, while for $\frac{1}{2} < H < 1$,

$$\begin{aligned}
|(M_H^+ f)(x)| &\leq C''_H \int_{|y-x|<1} dy\,|f(y)||y-x|^{H-\frac{3}{2}} + C''_H \int_{|y-x|>1} dy\,|f(y)||y-x|^{H-\frac{3}{2}} \\
&\leq C''_H \frac{2}{H-\frac{1}{2}} \sup_{x\in\mathbb{R}}|f(x)| + C''_H \int_{|y-x|>1} dy\,|f(y)||y-x|^{H-\frac{3}{2}},
\end{aligned}$$

for some constant $C''_H > 0$ also independent of f. Concerning the latter integral, observe that

$$\int_{|y-x|>1} dy\,|f(y)||y-x|^{H-\frac{3}{2}} \leq |f|\left(\int_{|y-x|>1} dy\,|y-x|^{2H-3}\right)^{1/2} = \frac{1}{\sqrt{1-H}}|f|,$$

leading to $\sup_{x\in\mathbb{R}}|(M_H^+ f)(x)| \le C_H''\left(\frac{2}{H-\frac{1}{2}}\sup_{x\in\mathbb{R}}|f(x)| + \frac{1}{\sqrt{1-H}}|f|\right)$. ■

We are now ready to state the main result on intersection local times L_H as well as on their subtracted counterparts $L_H^{(N)}$. For simplicity we shall use the notation

$$\Delta := \{(t_1,t_2) \in \mathbb{R}^2 : 0 < t_1 < t_2 < 1\}.$$

Theorem 7 *For any $H \in (0,1)$ and for any pair of integer numbers $d \ge 1$ and $N \ge 0$ such that $2N(H-1)+dH < 1$, the Bochner integral*

$$L_H^{(N)} := \int_\Delta d^2t\, \delta^{(N)}(\mathbf{B}_H(t_2) - \mathbf{B}_H(t_1))$$

is a Hida distribution.

Proof. To prove this result we shall again use Corollary 4 with respect to the Lebesgue measure on Δ. For this purpose let us denote the truncated exponential series by

$$\exp_N(x) := \sum_{n=N}^{\infty} \frac{x^n}{n!}.$$

It follows from (8) that the S-transform of $\delta^{(N)}$ is given by

$$S(\delta^{(N)}(\mathbf{B}_H(t_2) - \mathbf{B}_H(t_1)))(\mathbf{f}) \tag{11}$$
$$= \left(\frac{1}{\sqrt{2\pi}|t_2-t_1|^H}\right)^d \exp_N\left(-\frac{1}{2|t_2-t_1|^{2H}}\left|\int_{\mathbb{R}} dx\, \mathbf{f}(x)(M_H 1\!\!1_{[t_1,t_2]})(x)\right|^2\right),$$

which is a measurable function.

In order to check the boundedness condition, on $S_d(\mathbb{R})$ let us consider the norm $\|\cdot\|$ defined by

$$\|\mathbf{f}\| := \left(\sum_{i=1}^{d}\left(\sup_{x\in\mathbb{R}}|f_i(x)| + \sup_{x\in\mathbb{R}}|f_i'(x)| + |f_i|\right)^2\right)^{\frac{1}{2}}, \quad \mathbf{f} = (f_1,\dots,f_d) \in S_d(\mathbb{R}). \tag{12}$$

We observe that for $d = 1$ this norm reduces to a continuous norm on $S_1(\mathbb{R})$,

$$\|f\| = \sup_{x\in\mathbb{R}}|f(x)| + \sup_{x\in\mathbb{R}}|f'(x)| + |f|, \quad f \in S_1(\mathbb{R}),$$

which implies the continuity of the norm (12) for higher dimensions. By Lemma 6 we obtain

$$\left|\int_{\mathbb{R}} dx\, \mathbf{f}(x)(M_H 1\!\!1_{[t_1,t_2]})(x)\right|^2 \le C_H^2 |t_2-t_1|^2 \|\mathbf{f}\|^2,$$

and thus

$$\left|S(\delta^{(N)}(\mathbf{B}_H(t_2) - \mathbf{B}_H(t_1)))(z\mathbf{f})\right| \le \left(\frac{1}{\sqrt{2\pi}|t_2-t_1|^H}\right)^d \exp_N\left(\frac{C_H^2}{2}|z|^2|t_2-t_1|^{2-2H}\|\mathbf{f}\|^2\right),$$

for all $z \in \mathbb{C}$ and all $\mathbf{f} \in S_d(\mathbb{R})$. Estimating the function $\exp_N$ by

$$\exp_N\left(\frac{C_H^2}{2}|z|^2|t_2-t_1|^{2-2H}\|\mathbf{f}\|^2\right) \leq |t_2-t_1|^{2N(1-H)}e^{\frac{C_H^2}{2}|z|^2\|\mathbf{f}\|^2},$$

we then obtain

$$\left|S(\delta^{(N)}(\mathbf{B}_H(t_2)-\mathbf{B}_H(t_1)))(z\mathbf{f})\right| \leq \left(\frac{1}{\sqrt{2\pi}}\right)^d |t_2-t_1|^{2N(1-H)-dH}e^{\frac{C_H^2}{2}|z|^2\|\mathbf{f}\|^2},$$

where $|t_2-t_1|^{2N(1-H)-dH}$ is integrable on Δ if, and only if, $2N(1-H)-dH > -1$. The proof is completed by an application of Corollary 4. ■

As a consequence, one may derive the chaos expansion for the local times $L_H^{(N)}$.

Proposition 8 *Given a $H \in (0,1)$ and a pair of integer numbers $d \geq 1$ and $N \geq 0$ such that $2N(H-1)+dH < 1$, the kernel functions $F_{H,\mathbf{n}}$ are given by*

$$F_{H,2\mathbf{n}}(u_1,\ldots,u_{2n}) = \frac{1}{\mathbf{n}!}\left(\frac{1}{\sqrt{2\pi}}\right)^d\left(-\frac{1}{2}\right)^n\int_\Delta d^2t\,\frac{1}{|t_2-t_1|^{2Hn+dH}}\prod_{i=1}^{2n}(M_H 1\!\!1_{[t_1,t_2]})(u_i)$$

for each $\mathbf{n} \in \mathbb{N}^d$ such that $n \geq N$. All other kernel functions $F_{H,\mathbf{n}}$ are identically equal to zero.

Proof. By Corollary 4, the S-transform of the (truncated) local time $L_H^{(N)}$ is obtained by integrating (11) over Δ. Hence, given a $\mathbf{f} = (f_1,\ldots,f_d) \in S_d(\mathbb{R})$,

$$\begin{aligned}
SL_H^{(N)}(\mathbf{f}) &= \left(\frac{1}{\sqrt{2\pi}}\right)^d\int_\Delta d^2t\,\frac{1}{|t_2-t_1|^{dH}}\sum_{n=N}^\infty\frac{(-1)^n}{2^n|t_2-t_1|^{2Hn}}\times \\
&\quad\times\sum_{\substack{n_1,\cdots,n_d\\ n_1+\cdots+n_d=n}}\frac{1}{n_1!\cdots n_d!}\prod_{j=1}^d\left(\int_\mathbb{R}dx\,f_j(x)(M_H 1\!\!1_{[t_1,t_2]})(x)\right)^{2n_j} \\
&= \left(\frac{1}{\sqrt{2\pi}}\right)^d\int_\Delta d^2t\sum_{n=N}^\infty\left(-\frac{1}{2}\right)^n\frac{1}{|t_2-t_1|^{2Hn+dH}}\times \\
&\quad\times\sum_{\substack{n_1,\cdots,n_d\\ n_1+\cdots+n_d=n}}\frac{1}{\mathbf{n}!}\prod_{j=1}^d\left(\int_\mathbb{R}dx\,f_j(x)(M_H 1\!\!1_{[t_1,t_2]})(x)\right)^{2n_j}.
\end{aligned}$$

Comparing with the general form of the chaos expansion

$$L_H^{(N)} = \sum_\mathbf{n}\langle:\omega^{\otimes\mathbf{n}}:,F_{H,\mathbf{n}}\rangle,$$

one concludes that

$$F_{H,2\mathbf{n}}(u_1,\ldots,u_{2n}) = \frac{1}{\mathbf{n}!}\left(\frac{1}{\sqrt{2\pi}}\right)^d\left(-\frac{1}{2}\right)^n\int_\Delta d^2t\,\frac{1}{|t_2-t_1|^{2Hn+dH}}\prod_{i=1}^{2n}(M_H 1\!\!1_{[t_1,t_2]})(u_i)$$

for each $\mathbf{n} \in \mathbb{N}^d$ such that $n \geq N$, while all other kernels $F_{H,\mathbf{n}}$ vanish. ■

Theorem 7 shows that for $d = 1$ all intersection local times L_H are well-defined for all possible Hurst parameters H in $(0,1)$. For $d \geq 2$, intersection local times are well-defined only for $H < 1/d$. Informally speaking, for $H \geq 1/d$ with $d \geq 2$, the local times only become well-defined once subtracted the divergent terms. This "renormalization" procedure motivates the study of a regularization. As a computationally simple regularization we discuss

$$L_{H,\varepsilon} := \int_\Delta d^2t\, \delta_\varepsilon(\mathbf{B}_H(t_2) - \mathbf{B}_H(t_1)), \quad \varepsilon > 0,$$

where

$$\delta_\varepsilon(\mathbf{B}_H(t_2) - \mathbf{B}_H(t_1)) := \left(\frac{1}{\sqrt{2\pi\varepsilon}}\right)^d e^{-\frac{(\mathbf{B}_H(t_2)-\mathbf{B}_H(t_1))^2}{2\varepsilon}}.$$

Theorem 9 *Let $\varepsilon > 0$ be given. For all $H \in (0,1)$ and all dimensions $d \geq 1$ the functional $L_{H,\varepsilon}$ is a Hida distribution with kernel functions given by*

$$F_{H,\varepsilon,2\mathbf{n}}(u_1,\ldots,u_{2n}) = \frac{1}{\mathbf{n}!}\left(\frac{1}{\sqrt{2\pi}}\right)^d \left(-\frac{1}{2}\right)^n \int_\Delta d^2t \frac{1}{(\varepsilon + |t_2 - t_1|^{2H})^{n+\frac{d}{2}}} \prod_{i=1}^{2n} (M_H \mathbb{1}_{[t_1,t_2]})(u_i)$$

for each $\mathbf{n} = (n_1,\ldots,n_d) \in \mathbb{N}^d$, and $F_{H,\varepsilon,\mathbf{n}} \equiv 0$ if at least one of the n_i is an odd number. Moreover, if $2N(H-1) + dH < 1$, then when ε goes to zero the (truncated) functional $L_{H,\varepsilon}^{(N)}$ converges strongly in $(S)^*$ to the (truncated) local time $L_H^{(N)}$.

Proof. As before, the first part of the proof follows from the Corollary 4 with respect to the Lebesgue measure on Δ. By the definition of the S-transform, for all $\mathbf{f} \in S_d(\mathbb{R})$ one finds

$$S\delta_\varepsilon(\mathbf{B}_H(t_2) - \mathbf{B}_H(t_1))(\mathbf{f}) = \left(\frac{1}{\sqrt{2\pi(\varepsilon + |t_2 - t_1|^{2H})}}\right)^d e^{-\frac{1}{2(\varepsilon+|t_2-t_1|^{2H})}\left|\int_{\mathbb{R}} dx\, \mathbf{f}(x)(M_H \mathbf{1}_{[t_1,t_2]})(x)\right|^2},$$

which is measurable. Hence, by Lemma 6, for all $z \in \mathbb{C}$ and all $\mathbf{f} \in S_d(\mathbb{R})$ one has

$$|S\delta_\varepsilon(\mathbf{B}_H(t_2) - \mathbf{B}_H(t_1))(z\mathbf{f})| \leq \left(\frac{1}{\sqrt{2\pi(\varepsilon + |t_2 - t_1|^{2H})}}\right)^d e^{C_H^2|z|^2 \frac{|t_2-t_1|^2}{2(\varepsilon+|t_2-t_1|^{2H})}\|\mathbf{f}\|^2},$$

with $\frac{|t_2-t_1|^2}{2(\varepsilon+|t_2-t_1|^{2H})}$ bounded on Δ and $(\varepsilon+|t_2-t_1|^{2H})^{-\frac{d}{2}}$ integrable on Δ. By Corollary 4, one may then conclude that $L_{H,\varepsilon}\in(S)^*$ and, moreover, for every $\mathbf{f}=(f_1,\ldots,f_n)\in S_d(\mathbb{R})$,

$$\begin{aligned}
SL_{H,\varepsilon}(\mathbf{f}) &= \int_\Delta d^2t\, S\delta_\varepsilon(\mathbf{B}_H(t_2)-\mathbf{B}_H(t_1))(\mathbf{f})\\
&= \left(\frac{1}{\sqrt{2\pi}}\right)^d \int_\Delta d^2t \frac{1}{(\varepsilon+|t_2-t_1|^{2H})^{\frac{d}{2}}}\sum_{n=0}^{\infty}\frac{(-1)^n}{2^n(\varepsilon+|t_2-t_1|^{2H})^n}\times\\
&\quad\times \sum_{\substack{n_1,\cdots,n_d\\ n_1+\cdots+n_d=n}}\frac{1}{n_1!\cdots n_d!}\prod_{j=1}^d\left(\int_{\mathbb{R}}dx\, f_j(x)(M_H 1\!\!1_{[t_1,t_2]})(x)\right)^{2n_j}\\
&= \left(\frac{1}{\sqrt{2\pi}}\right)^d \int_\Delta d^2t\sum_{n=0}^{\infty}\left(-\frac{1}{2}\right)^n\frac{1}{(\varepsilon+|t_2-t_1|^{2H})^{n+\frac{d}{2}}}\times\\
&\quad\times \sum_{\substack{n_1,\cdots,n_d\\ n_1+\cdots+n_d=n}}\frac{1}{\mathbf{n}!}\prod_{j=1}^d\left(\int_{\mathbb{R}}dx\, f_j(x)(M_H 1\!\!1_{[t_1,t_2]})(x)\right)^{2n_j}.
\end{aligned}$$

As in the proof of Proposition 8, it follows from the latter expression that the kernels $F_{H,\varepsilon,\mathbf{n}}$ appearing in the chaos expansion $L_{H,\varepsilon}=\sum_{\mathbf{n}}\langle :\omega^{\otimes\mathbf{n}}:,F_{H,\varepsilon,\mathbf{n}}\rangle$ vanish if at least one of the n_i in $\mathbf{n}=(n_1,\ldots,n_d)$ is an odd number, otherwise they are given by

$$F_{H,\varepsilon,2\mathbf{n}}(u_1,\ldots,u_{2n})=\frac{1}{\mathbf{n}!}\left(\frac{1}{\sqrt{2\pi}}\right)^d\left(-\frac{1}{2}\right)^n\int_\Delta d^2t\frac{1}{(\varepsilon+|t_2-t_1|^{2H})^{n+\frac{d}{2}}}\prod_{i=1}^{2n}(M_H 1\!\!1_{[t_1,t_2]})(u_i).$$

To complete the proof amounts to check the convergence. For this purpose we shall use Corollary 3. Since

$$SL_{H,\varepsilon}^{(N)}(\mathbf{f})=\int_\Delta d^2t\, S\delta_\varepsilon^{(N)}(\mathbf{B}_H(t_2)-\mathbf{B}_H(t_1))(\mathbf{f}),$$

for all $z\in\mathbb{C}$ and all $\mathbf{f}\in S_d(\mathbb{R})$ we have

$$\begin{aligned}
\left|SL_{H,\varepsilon}^{(N)}(z\mathbf{f})\right| &\le \int_\Delta d^2t\,|S\delta_\varepsilon(\mathbf{B}_H(t_2)-\mathbf{B}_H(t_1))(z\mathbf{f})|\\
&\le \left(\frac{1}{\sqrt{2\pi\varepsilon}}\right)^d\int_\Delta d^2t\, e^{\frac{C_H^2}{2\varepsilon}|z|^2|t_2-t_1|^2\|\mathbf{f}\|^2}\le\left(\frac{1}{\sqrt{2\pi\varepsilon}}\right)^d e^{\frac{C_H^2}{2\varepsilon}|z|^2\|\mathbf{f}\|^2},
\end{aligned}$$

showing the boundedness condition. In addition, similar computations used to prove Theorem 7 yield for all $(t_1,t_2)\in\Delta$

$$\begin{aligned}
\left|S\delta_\varepsilon^{(N)}(\mathbf{B}_H(t_2)-\mathbf{B}_H(t_1))(\mathbf{f})\right| &\le \left(\frac{1}{\sqrt{2\pi}|t_2-t_1|^H}\right)^d\exp_N\left(\frac{C_H^2}{2}|t_2-t_1|^{2-2H}\|\mathbf{f}\|^2\right)\\
&\le \left(\frac{1}{\sqrt{2\pi}}\right)^d|t_2-t_1|^{2N(1-H)-dH}e^{\frac{C_H^2}{2}\|\mathbf{f}\|^2},
\end{aligned}$$

which allows the use of the Lebesgue dominated convergence theorem to infer the other condition needed for the application of Corollary 3. ■

ACKNOWLEDGMENTS

M.J.O. and J.L.S. would like to express their gratitude for the splendid hospitality of our colleagues and friends Victoria Bernido and Christopher Bernido during a very pleasant stay at Jagna during the 5th Jagna International Workshop. Financial support of the FCT through POCI, PDCT, and PTDC projects is gratefully acknowledged.

REFERENCES

1. C. Bender. An Itô formula for generalized functionals of a fractional Brownian motion with arbitrary Hurst parameter. *Stochastic Processes Appl.*, 104:81–106, 2003.
2. Yu. M. Berezansky and Yu. G. Kondratiev. *Spectral Methods in Infinite-Dimensional Analysis*. Naukova Dumka, Kiev, 1988. (in Russian). English translation, Kluwer Academic Publishers, Dordrecht, 1995.
3. M. de Faria, C. Drumond, and L. Streit. The renormalization of self-intersection local times. I. The chaos expansion. *Infin. Dimens. Anal. Quantum Probab. Relat. Top.*, 3(2):223–236, 2000.
4. M. de Faria, T. Hida, L. Streit, and H. Watanabe. Intersection local times as generalized white noise functionals. *Acta Appl. Math.*, 46:351–362, 1997.
5. M. Gradinaru, F. Russo, and P. Vallois. Generalized covariations, local time and Stratonovich Itô's formula for fractional Brownian motion with Hurst index $H \geq \frac{1}{4}$. *Ann. Probab.*, 31(4):1772–1820, 2003.
6. T. Hida. *Analysis of Brownian Functionals*, volume 13 of *Carleton Mathematical Lecture Notes*. Carleton University, Ottawa, 1975.
7. T. Hida, H. H. Kuo, J. Potthoff, and L. Streit. *White Noise. An Infinite Dimensional Calculus*. Kluwer Academic Publishers, Dordrecht, 1993.
8. Y. Hu and D. Nualart. Renormalized self-intersection local time for fractional Brownian motion. *Ann. Probab.*, 33:948–983, 2005.
9. Y. Hu and D. Nualart. Regularity of renormalized self-intersection local time for fractional Brownian motion. *Commun. Inf. Syst.*, 7(1):21–30, 2007.
10. H. Holden, B. Øksendal, J. Ubøe, and T. Zhang. *Stochastic Partial Differential Equations: A Modeling, White Noise Functional Approach*. Birkhäuser, Boston, Basel, and Berlin, 1996.
11. Yu. G. Kondratiev, P. Leukert, J. Potthoff, L. Streit, and W. Westerkamp. Generalized functionals in Gaussian spaces: The characterization theorem revisited. *J. Funct. Anal.*, 141:301–318, 1996.
12. Yu. G. Kondratiev. Spaces of Test and Generalized Functions of an Infinite Number of Variables. Master's thesis, University of Kiev, 1975.
13. I. Kubo and S. Takenaka. Calculus on Gaussian white noise I. *Proc. Japan Acad. Ser. A Math. Sci.*, 56:376–380, 1980.
14. I. Kubo and S. Takenaka. Calculus on Gaussian white noise II. *Proc. Japan Acad. Ser. A Math. Sci.*, 56:411–416, 1980.
15. H. H. Kuo. *White Noise Distribution Theory*. CRC Press, Boca Raton, New York, London, and Tokyo, 1996.
16. D. Nualart and S. Ortiz-Latorre. Intersection local time for two independent fractional Brownian motions. *J. Theoret. Probab.*, 20(4):759–767, 2007.
17. N. Obata. *White Noise Calculus and Fock Space*, volume 1577 of *LNM*. Springer Verlag, Berlin, Heidelberg, and New York, 1994.
18. J. Potthoff and L. Streit. A characterization of Hida distributions. *J. Funct. Anal.*, 101:212–229, 1991.
19. V. Pipiras and M. S. Taqqu. Integration questions related to fractional Brownian motion. *Probab. Theory Related Fields*, 118:251–291, 2000.
20. A. Rezgui. The renormalization of self intersection local times of fractional Brownian motion. *Int. Math. Forum*, 2(44):2161–2178, 2007.
21. J. Rosen. The intersection local time of fractional Brownian motion in the plane. *J. Multivar. Anal.*, 23:37–46, 1987.

Time Development of Explosion of Stochastic Process with Repulsive Drift of Polynomial Growth

Toru Nakamura*, Hiroshi Ezawa†, Keiji Watanabe¶

*Department of Mathematics, Sundai Preparatory School, Kanda-Surugadai, Chiyoda-ku, Tokyo 101-0062, Japan
†Department of Physics, Gakushuin University, Mejiro, Toshima-ku, Tokyo 171-8588, Japan[1]
¶Department of Physics, Meisei University, Hino, Tokyo 191-8506, Japan[2]

Abstract. For a diffusion process $X(t)$ representing the momentum of a particle subject to an external force $f(x)$ and a random force $dB(t)/dt$ in one-dimensional space, almost all sample paths explode to infinity in finite times if $f(x)$ grows repulsively to infinity faster than linear as $|x| \to \infty$, so that the survival probability $P(t)$ of the particle by time t decreases to zero as time passes. It is shown that $P(t)$ decays at least exponentially in time and the rate of the exponential decay is strictly positive and equal to the lowest eigenvalue of a Hamiltonian of an imaginary-time Schrödinger equation.

Keywords: Explosion, Time Development, WKB Method.
PACS: 02.30.Tb, 02.50.Ey

INTRODUCTION

We consider a particle subject to an external force $f(X(t))$ and a random force or white noise $dB(t)/dt$ in one-dimensional space. Here, $X(t)$ denotes the momentum of the particle at time t, so that it satisfies the stochastic differential equation,

$$dX(t) = f\big(X(t)\big)\,dt + dB(t), \tag{1}$$

representing Newton's equation of motion. We assume that the force $f(x)$ acts on the particle repulsively for large $|x|$, that is, the signs of $f(x)$ and x are the

[1] Mailing address: 1-25-5 Honkomagome, Bunkyo-ku, Tokyo 113-0021, Japan.
[2] Mailing address: 5-36-2 Akazutsumi, Setagaya-ku, Tokyo 156-0044, Japan.

CP1021, *Stochastic and Quantum Dynamics of Biomolecular Systems, Proceedings of the 5th Jagna International Workshop,* edited by C. C. Bernido and M. V. Carpio-Bernido

same for large $|x|$, and that $f(x)$ goes to infinity as $|x| \to \infty$ with a polynomial growth of degree greater than 1. We set the following conditions for $f(x)$:

(C1) $f(x) \in \mathsf{C}^2(\mathbb{R})$, namely, $f(x)$ is differentiable twice and its second derivative is continuous.

(C2) $f(x)$ can be expanded as

$$f(x) = \sum_{k=-\infty}^{m} c_k x^k \quad (x > x_0)$$

and

$$f(x) = \sum_{k=-\infty}^{m'-1} c'_k x^k + c'_{m'} |x|^{m'} \quad (x < -x_0)$$

for some constant $x_0 > 0$. The coefficients of the highest order are such that $c_m > 0$ and $c'_{m'} < 0$, and the degrees m and m' for each side of x are both greater than 1.

We have assumed for simplicity that $c_m > 0$ and $c'_{m'} < 0$ in (C2) guaranteeing that the force $f(x)$ is repulsive on both sides of x, positive and negative. In actual fact, it is proved that Theorem 1 and 2 below hold if $f(x)$ is repulsive at least on one side of x (see [1]).

Under these conditions, it is known that the process $X(t)$ almost surely explodes to infinity in finite time, and the time development of the explosion has been studied in probability theory using speed and scale [2,3,4]. In the present paper, we shall show in Theorem 2 below that the time development relates to the lowest eigenvalue of the Hamiltonian of an imaginary-time Schrödinger equation.

We define the probability density of the particle momentum x at time t by

$$\phi(t,x) := \int_{\mathcal{P}_{t,x}} d\mu(X_\omega), \tag{2}$$

where the domain of the integration is

$$\mathcal{P}_{t,x} = \{X_\omega \mid X_\omega(t) = x\},$$

and the measure μ is defined over a space of the sample paths X_ω, for example, by making use of nonstandard analysis [5]. Then, using the properties of the measure, it is proved that $\phi(t,x)$ satisfies the forward Fokker-Planck equation (FP-equation),

$$\frac{\partial}{\partial t}\phi(t,x) = D\frac{\partial^2}{\partial x^2}\phi(t,x) - \frac{\partial}{\partial x}\{f(x)\phi(t,x)\}. \tag{3}$$

It is also proved that those paths are negligible which once reach an infinite momentum and return to finite x at time t, that is, such paths do not contribute to the probability density $\phi(t,x)$ defined by (2) (see [5] for the proofs).

Therefore, we define the survival probability of the particle by time t by

$$P(t) := \int_{-\infty}^{\infty} \phi(t,x)dx, \tag{4}$$

and use WKB method to show that

$$P(t) \leq ce^{-E_0 t} \quad (t > t_0)$$

for some positive constants c, t_0 and E_0, where $t_0 > 0$ can be chosen arbitrarily and $c > 0$ depends on the choice of t_0. The rate E_0 of the exponential decay will be shown to be (i) strictly positive and (ii) the lowest eigenvalue of the Hamiltonian having the potential $V(x)$ defined by

$$V(x) := \frac{f(x)^2}{4D} + \frac{f'(x)}{2}. \tag{5}$$

We note that the condition (C2) implies that $V(x)$ is bounded from below and that $V(x)$ goes to infinity as $x \to \infty$.

FROM FP-EQUATION TO SCHRÖDINGER EQUATION

Since $\phi(0,x)$ denotes the probability density of the particle momentum x at time 0, we set the following condition for the initial data:

(C3) $\phi(0,x) \in \mathsf{C}_0^2(\mathbb{R})$, that is, $\phi(0,x) \in \mathsf{C}^2(\mathbb{R})$ having a compact support. Moreover, it satisfies

$$\phi(0,x) \geq 0 \quad \text{and} \quad \int_{-\infty}^{\infty} \phi(0,x)dx = 1.$$

We change the variable $\phi(t,x)$ to

$$\psi(t,x) := \phi(t,x)\exp\left[-U(x)/(2D)\right]$$

using

$$U(x) := \int_0^x f(y)dy.$$

Then, $\psi(t,x)$ satisfies the imaginary-time Schrödinger equation,

$$-\frac{\partial\psi(t,x)}{\partial t} = \mathcal{H}\psi(t,x), \tag{6}$$

where the Hamiltonian is given by

$$\mathcal{H} := -D\frac{\partial^2}{\partial x^2} + V(x) \tag{7}$$

with the potential $V(x)$ in (5) satisfying $\lim_{x\to\pm\infty} V(x) = \infty$ as noted before. It is known that the Hamiltonian with such potential can be extended to a self-adjoint operator on a domain dense in $\mathsf{L}^2(\mathbb{R})$ having a purely discrete spectrum with a complete orthonormal system $\{u_n(x) \in \mathsf{L}^2(\mathbb{R}) \mid n = 0, 1, \cdots\}$ of eigenfunctions,

$$\mathcal{H}u_n(x) = E_n u_n(x) \quad (E_0 < \cdots < E_n < \cdots \to \infty). \tag{8}$$

Further, the n-th eigenfunction $u_n(x)$ has exactly n zeros (see [6]).

THEOREMS AND THEIR PROOFS

Despite the term $f'(x)/2$ in (5) which may take negative values, the Hamiltonian $\mathcal{H}$ is a positive operator.

Theorem 1 *The lowest eigenvalue of $\mathcal{H}$ in (7) is strictly positive, that is,*

$$E_0 > 0. \tag{9}$$

Proof: Let

$$w_0(x) := u_0(x)\exp\Big[\frac{U(x)}{2D}\Big]$$

which can be set to be positive for all x because $u_0(x)$ is the eigenfunction for the lowest eigenvalue and has no zeros. Since the characteristic equation (8) for $n = 0$ becomes

$$E_0 w_0(x) = -D\frac{d^2 w_0(x)}{dx^2} + \frac{d}{dx}\{f(x)w_0(x)\},$$

we obtain

$$E_0 \int_{-\infty}^{\infty} w_0(x)dx \;=\; -D\left[\frac{dw_0(x)}{dx}\right]_{-\infty}^{\infty} + \Big[f(x)w_0(x)\Big]_{-\infty}^{\infty}. \tag{10}$$

To calculate the right-hand side, we use the WKB approximation,

$$u_0^{\mathrm{app}}(x) = \pm\frac{N_\pm}{f(x)}\exp\Big[-\frac{U(x)}{2D}\Big] \quad (x \to \pm\infty)$$

with the double signs being read in the same order, to replace $w_0(x)$ in (10) by

$$w_0^{\mathrm{app}}(x) = u_0^{\mathrm{app}}(x)\exp\Big[\frac{U(x)}{2D}\Big] = \pm\frac{N_\pm}{f(x)} \tag{11}$$

and $dw_0(x)/dx$ by

$$\frac{dw_0^{\rm app}(x)}{dx} = \mp\frac{N_\pm f'(x)}{f(x)^2}. \tag{12}$$

We skip the proof of the inequalities,

$$\left|u_0(x) - u_0^{\rm app}(x)\right| < {\rm const.}k^{-1}\left|u_0^{\rm app}(x)\right| \quad (|x| > x_1)$$

and

$$\left|\frac{du_0(x)}{dx} - \frac{du_0^{\rm app}(x)}{dx}\right| < {\rm const.}k^{-1}\left|\frac{du_0^{\rm app}(x)}{dx}\right| \quad (|x| > x_1)$$

for some constant $x_1 > 0$, where k is introduced as $x = kz$ so that we take $k \to \infty$ instead of $x \to \infty$ and $N_\pm$ are constants determined by the normalization of $u_0^{\rm app}(x)$ (see [1] or [7] for the proof). Then, the replacements of $w_0(x)$ and $dw_0(x)/dx$ in (10) respectively by (11) and (12) are justified using these inequalities. Hence

$$E_0\int_{-\infty}^{\infty} w_0(x)dx = -D\left[\mp\frac{N_{0\pm} f'(x)}{f(x)^2}\right]_{-\infty}^{\infty} + \Big[\pm N_\pm\Big]_{-\infty}^{\infty} = N_+ + N_- > 0,$$

so that (9) holds because $w_0(x) > 0$ for all x. ∎

The next theorem says that E_0 in Theorem 1 is the rate of the exponential decay of the survival probability.

Theorem 2 *The survival probability $P(t)$ defined by (4) satisfies*

$$P(t) \le ce^{-E_0 t} \quad (t > t_0), \tag{13}$$

where the constant c depends on $t_0 > 0$ chosen arbitrarily.

Proof: Since the set of the eigenfunctions, $\{u_n(x) \in \mathsf{L}^2(\mathbb{R}) \mid n = 0, 1, \cdots\}$, is a complete orthonormal system, $\psi(0,x)$ which belongs to $\mathsf{C}_0^2(\mathbb{R})$ by (C3) can be expanded in a series of the eigenfunctions,

$$\psi(0,x) = \sum_{n=0}^{\infty} c_n u_n(x) \quad \text{with} \quad c_n = \left\langle \psi(0,x),\, u_n(x)\right\rangle_2.$$

Then,

$$\psi(t,x) = e^{-\mathcal{H}t}\psi(0,x) = \sum_{n=0}^{\infty} c_n e^{-E_n t} u_n(x),$$

so that

$$
\begin{aligned}
P(t) =& e^{-E_0 t} \int_{-\infty}^{\infty} \sum_{n=0}^{\infty} e^{-(E_n - E_0)t} c_n u_n(x) \exp\Big[\frac{U(x)}{2D}\Big] dx \\
\leq& e^{-E_0 t} \int_{-\infty}^{\infty} \sum_{n=0}^{\infty} e^{-(E_n - E_0)t_0} |c_n|\, |u_n(x)| \exp\Big[\frac{U(x)}{2D}\Big] dx
\end{aligned} \tag{14}
$$

for arbitrary $t_0 > 0$. If the integral of the infinite sum in the last line converges, we can write (14) as

$$
P(t) \leq c e^{-E_0 t} \quad (t > t_0)
$$

by putting

$$
c = \int_{-\infty}^{\infty} \sum_{n=0}^{\infty} e^{-(E_n - E_0)t_0} |c_n|\, |u_n(x)| \exp\Big[\frac{U(x)}{2D}\Big] dx. \tag{15}
$$

Let us prove the convergence of the integral of the infinite sum on the right-hand side of (15). Interchanging the integration and the summation by Fatou's lemma and then applying the Cauchy-Schwarz inequality, we obtain

$$
c^2 \leq \|\psi(0,x)\|_2^2 \sum_{n=0}^{\infty} e^{-2(E_n - E_0)t_0} \Big(\int_{-\infty}^{\infty} |u_n(x)| \exp\Big[\frac{U(x)}{2D}\Big] dx \Big)^2. \tag{16}
$$

We again apply the WKB method to obtain the approximation of $u_n(x)$. However, we cannot use the large parameter k in the proof of Theorem 1 because the approximation of $u_n(x)$ in the present case should be taken for all x, not only for large x in the case of Theorem 1. Instead, we put

$$
E_n = D k_n^2
$$

and use k_n as the large parameter, because the convergence of the infinite sum depends on large n and hence on large k_n. It should be noticed that the WKB method can be applied only to x apart from the turning points q_n defined by

$$
V(q_n) = E_n.
$$

Therefore, we divide the whole axis $\mathbb{R}$ into seven intervals as shown in Fig. 1, where $V(x)$ is assumed to be an even function for large $|x|$ for simplicity. The magnitude of q_n satisfies

$$
q_n = \mathcal{O}(k_n^{1/m})
$$

with m in (C2), and the width h of the intervals I_j for $j = 2, 3, 5, 6$ is chosen such that

$$
h = \mathcal{O}(q_n^{\nu}) = \mathcal{O}(k_n^{\nu/m})
$$

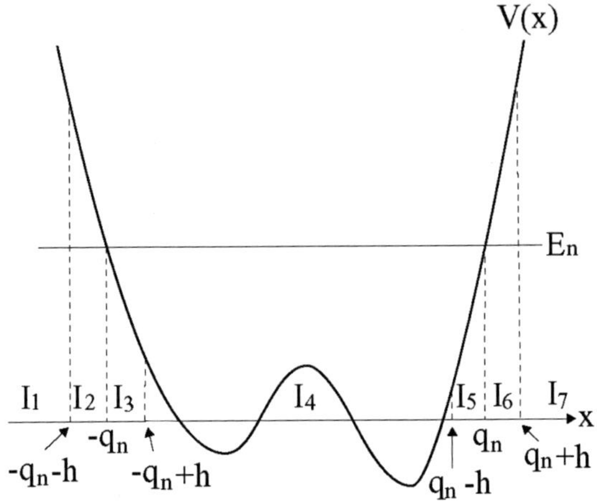

Figure 1: The seven intervals I_j

for some $0 < \nu < 1$ to be determined later. We apply WKB method for the intervals I_1, I_4 and I_7. For the rest of the intervals adjacent to the turning points $\pm q_n$, we approximate the potential as

$$V(x) \simeq V(\pm q_n) + V'(\pm q_n)\{x - (\pm q_n)\}.$$

Hereafter, we shall write the results only, leaving the details to [1] or [7]. The results show that the approximate eigenfunction is given for each interval as $u_n^{\mathrm{app}}(x) = u_{n,j}^{\mathrm{app}}(x)$ for $x \in I_j$, where

$$u_{n,j}^{\mathrm{app}}(\xi_j) = p_n(x)^{-1/4} \times \begin{cases} A_j e^{-k_n \xi_j} & (j = 1, 7) \\ A_j e^{ik_n \xi_j} + A'_j e^{-ik_n \xi_j} & (j = 4) \\ A_j H_{1/3}^{(1)}(ik_n \xi_j) & (j = 2,\ 6) \\ A_j H_{1/3}^{(1)}(k_n \xi_j) + A'_j H_{1/3}^{(2)}(k_n \xi_j) & (j = 3, 5) \end{cases} . \tag{17}$$

Here, $H_{1/3}^{(\iota)}(z)$ $(\iota = 1, 2)$ are the Hankel functions, $p_n(x)$ is defined by

$$p_n(x) := \begin{cases} V(x)/E_n - 1 & (|x| > q_n) \\ 1 - V(x)/E_n & (|x| < q_n) \end{cases}$$

so that $p_n(x) > 0$, and

$$\xi_j := \int_{\Xi_j} p_n(x')^{1/2} dx'.$$

The domain Ξ_j of the integration is $[x,\ -q_n]$ for $j = 1, 2$, $[-q_n,\ x]$ for $j = 3, 4$, $[x,\ q_n]$ for $j = 5$ and $[q_n,\ x]$ for $j = 6, 7$ so that $\xi_j > 0$ for all j.

The factors A_j and A'_j are determined so that $u^{\mathrm{app}}_{n,j}(x)$ for each interval should be connected smoothly one by one to obtain the estimate,

$$|A_j| < \mathrm{const.} k_n^{1/2-(1+\nu)/(4m)}. \tag{18}$$

The explicit values of A_j and A'_j will be written in Remark 1 below.

To prove the convergence of the infinite series on the right-hand side of (16) with $u_n(x)$ replaced by $u^{\mathrm{app}}_n(x)$, it suffices to prove the convergence of

$$\sum_{n=0}^{\infty} e^{-2(E_n-E_0)t_0}\Big(\int_{I_j} \big|u^{\mathrm{app}}_{n,j}(x)\big| \exp\Big[\frac{U(x)}{2D}\Big] dx\Big)^2 \tag{19}$$

for each $j = 1, \cdots, 7$ because of the inequality

$$\sum_{n=0}^{\infty} a_n(b_{n,1} + \cdots + b_{n,7})^2 \le 7 \sum_{n=0}^{\infty} a_n\big({b_{n,1}}^2 + \cdots + {b_{n,7}}^2\big) \quad (a_n > 0)$$

where we denote

$$a_n = e^{-2(E_n-E_0)t_0} \quad \text{and} \quad b_{n,j} = \int_{I_j} \big|u^{\mathrm{app}}_{n,j}(x)\big| \exp\Big[\frac{U(x)}{2D}\Big] dx.$$

We shall give the outline of the proof of the convergence of (19), leaving the details again to [1] or [7].

For $j = 7$ where $I_7 = (q_n + h, \infty)$ and

$$u^{\mathrm{app}}_{n,7}(x) = A_7 p_n(x)^{-1/4} e^{-k_n \xi_7}, \tag{20}$$

$x \in I_7$ is sufficiently large to approximate $p_n(x)^{1/2} = (V(x)/E_n - 1)^{1/2}$ as

$$\begin{aligned} p_n(x)^{1/2} &= \Big\{\frac{f(x)^2}{4D^2k_n^2}\Big(1 + \frac{2Df'(x)}{f(x)^2}\Big)\Big\}^{1/2} \\ &\simeq \frac{f(x)}{2Dk_n}\Big(1 + \frac{Df'(x)}{f(x)^2}\Big) = \frac{f(x)}{2Dk_n} + \frac{f'(x)}{2k_n f(x)}, \end{aligned} \tag{21}$$

and hence

$$k_n \xi_7 \simeq \frac{U(x)}{2D} + \frac{1}{2}\log f(x).$$

Actually, it is proved that

$$-k_n \xi_7 < -\frac{U(x)}{2D} - \frac{1}{2}\log f(x) + \mathrm{const.} k_n^{1+(1/m)},$$

so that

$$e^{-k_n \xi_7} < \frac{1}{f(x)^{1/2}} \exp\Big[-\frac{U(x)}{2D} + \mathrm{const.} k_n^{1+(1/m)}\Big]. \tag{22}$$

For $p_n(x)^{-1/4}$, we use only the first term on the rightmost-hand side of (21) to have

$$p_n(x)^{-1/4} < \text{const.}\Big(\frac{k_n}{f(x)}\Big)^{1/2}. \tag{23}$$

Putting (18), (22) and (23) into (20), we obtain

$$\big|u_{n,7}^{\text{app}}(x)\big| \exp\Big[\frac{U(x)}{2D}\Big] < \text{const.}\frac{k_n^{1-(1+\nu)/(4m)}}{f(x)} \exp\big[\text{const.}k_n^{1+(1/m)}\big]$$

whose right-hand side is integrable over $[q_n + h, \infty)$ for $f(x)$ is assumed to grow faster than linear as $x \to \infty$. Thus the integral in (19) for $j = 7$ is finite and satisfies

$$\Big(\int_{I_7} \big|u_{n,7}^{\text{app}}(x)\big| \exp\Big[\frac{U(x)}{2D}\Big] dx\Big)^2 < \text{const.}k_n^{\sigma} \exp\big[\text{const.}k_n^{1+(1/m)}\big] \tag{24}$$

for some constant σ. Since $1 + (1/m) < 2$ by the assumption $m > 1$ in (C2),

$$e^{-2(E_n - E_0)t_0} = \text{const.} \exp\big[-2t_0 D k_n^2\big]$$

decreases to 0 stronger than the growth of the right-hand side of (24) as $k_n \to \infty$, so that the infinite sum (19) for $j = 7$ converges.

The convergence for other j is proved similarly, and hence the convergence of the infinite sum in (16) with $u_n(x)$ replaced by $u_n^{\text{app}}(x)$ is proved.

Finally, it is proved that the error from the replacement of $u_n(x)$ in (16) by $u_n^{\text{app}}(x)$ is negligible (see [1], [7] for the proof), for example for $j = 7$, using

$$\big|u_n(x) - u_{n,7}^{\text{app}}(x)\big| < \text{const.}k_n^{-\beta}\big|u_{n,7}^{\text{app}}(x)\big| \quad (x \in I_7) \tag{25}$$

with $\beta = (9\nu - 7 + 2m)/(2m)$ which is positive if we choose

$$\frac{7 - 2m}{9} < \nu < 1.$$

Thus, the convergence of the integral of the infinite sum in (15) is proved, and the proof of Theorem 2 is completed. ∎

Remark 1 By connecting $u_{n,j}^{\text{app}}(x)$ for each interval smoothly one by one, the normalization factors A_j and A'_j are determined to give

$$\begin{aligned}
A_2 &= \Big(\frac{\pi i k_n}{2}\Big)^{1/2} e^{5\pi i/12} A_1, & & \\
A_3 &= -iA_2, & A'_3 &= -e^{\pi i/6} A_2, \\
A_4 &= \Big(\frac{2}{\pi k_n}\Big)^{1/2} e^{-5\pi i/12} A_3, & A'_4 &= \Big(\frac{2}{\pi k_n}\Big)^{1/2} e^{5\pi i/12} A'_3,
\end{aligned}$$

and

$$A_6 = \left(\frac{\pi i k_n}{2}\right)^{1/2} e^{5\pi i/12} A_7,$$
$$A_5 = -i A_6, \qquad A'_5 = -e^{\pi i/6} A_6,$$
$$A_4 = e^{-ik_n\alpha_n}\left(\frac{2}{\pi k_n}\right)^{1/2} e^{5\pi i/12} A'_5, \quad A'_4 = e^{ik_n\alpha_n}\left(\frac{2}{\pi k_n}\right)^{1/2} e^{-5\pi i/12} A_5,$$

where

$$\alpha_n := \int_{-q_n}^{q_n} p_n(x)^{1/2} dx.$$

Equating the results of A'_4/A_4 from these two determinations, we obtain

$$k_n \alpha_n = \left(n + \frac{1}{2}\right)\pi,$$

or the Sommerfeld quantum condition,

$$\frac{1}{\sqrt{D}} \int_{-q_n}^{q_n} \{E_n - V(x)\}^{1/2} dx = \left(n + \frac{1}{2}\right)\pi.$$

This condition determines the energy eigenvalues. For example, if $V(x) = ax^{2m}$ then

$$E_n \simeq \left(ac^{2m}\right)^{1/(m+1)} \left(n + \frac{1}{2}\right)^{(2m)/(m+1)}, \quad c = \frac{\pi m \sqrt{D}}{B\left(\frac{3}{2}, \frac{1}{2m}\right)} \tag{26}$$

for large n, where $B(p, q)$ is the Beta function.

CONCLUDING REMARK

In this paper, our work is restricted to the process in one-dimension, and the extension to the higher dimensions is left to further study. It is interesting to find how large the repulsive domain of $f(x)$ should be in order that the process has the same features as those in one dimension. Do they hold true even when $f(x)$ is repulsive in a very small solid angle? How are they when the repulsive domain is cylindrical?

We would like to dedicate this paper to Professors Frederik Wiegel and Ludwig Streit on the occasion of their 70th birthday with respect.

REFERENCES

1. H. Ezawa, T. Nakamura and K. Watanabe, Time Development of Explosion and Path-Space Measure for Diffusion Process with Drift of Polynomial Growth, In preparation.

2. W. Feller, The Parabolic Differential Equations and the Associated Semigroups of Transformations, Ann. Math. **55**, 468–519 (1952).

3. H. P. McKean, Jr., *Stochastic Integrals*, Academic Press, New York and London (1969). §3.6.

4. K. Ito and H. P. McKean, Jr., *Diffusion Processes and their Sample Paths, Second Printing Corrected*, Springer-Verlag Berlin Heidelberg New York (1991).

5. T. Nakamura, H. Ezawa and K. Watanabe, Path-Space Measure for Diffusion Process with Drift of Polynomial Growth, In preparation.

6. F. A. Berezin and M. A. Shubin, *The Schrödinger equation*, Mathematics and Its Application vol. 66, Kluwer Acad. Publ. (1991).

7. H. Ezawa, T. Nakamura and K. Watanabe, Convergent WKB Series – How Can It be ?, This Proceedings.

Convergent WKB Series - How Can It be ?

Hiroshi Ezawa*, Toru Nakamura† and Keiji Watanabe**

*Department of Physics, Gakushuin University, Mejiro, Toshima-ku, Tokyo 171-8588, Japan[1]
†Department of Mathematics, Sundai Preparatory School, Kanda-Surugadai, Chiyoda-ku, Tokyo 101-0062, Japan
**Department of Physics, Meisei University, Hino-shi, Tokyo 191-8506, Japan[2]

Abstract. Schrödinger equation for a polynomial potential with the highest order term having an even power and a positive coefficient is solved for high eigenvalues E_n in two different ways after Liouville transformation, (a) converting the differential equation into integral equation and solving it iteratively and (b) by the WKB method. While the series solution in powers of $1/\sqrt{E_n}$ from (b) is known to diverge, we show that the one from (a) converges. We show then that asymptotic re-expansion of the convergent series from (a) agrees with the divergent series from (b). Actually, we have been able to show the agreement only up to order $(1/\sqrt{E_n})^5$, but we believe that it holds to all orders. If this is true, the divergent WKB series can be reorganized into a convergent series, which is in fact obtained by the method of iteration (a).

Keywords: WKB Series, Asymptotic Expansion, Convergence of Perturbation Series
PACS: 02.30.Hq, 02.30.Lt. 0230.Mv.

INTRODUCTION

In the preceding paper, "Time Development of Explosion of Diffusion Process with Drift of Polynomial Growth" [1], we solved the Schrödinger equation,

$$\mathscr{H} u_n(x) = E_n u_n(x), \quad \mathscr{H} := -D\frac{d^2}{dx^2} + V(x) \tag{1}$$

for large E_n, by Liouville transformation [2],[3]

$$\xi := \int^x p(x')^{1/2} dx', \qquad \Lambda(\xi) := p(x)^{1/4} u(x), \quad \left(p(x) := \frac{V(x)}{E} - 1\right), \tag{2}$$

$$\left(\frac{d^2}{d\xi^2} \mp k^2\right) \Lambda(\xi) = Q(\xi)\Lambda(\xi), \qquad \left(k^2 := \frac{E}{D}\right), \tag{3}$$

in iterative form,

$$\Lambda(\xi) = \Lambda^{(0)} + \Lambda^{(1)}(\xi) + \Lambda^{(2)}(\xi) + \cdots, \tag{4}$$

where, for

[1] Mailing address: 1-25-5 Honkomagome, Bunkyo-ku, Tokyo 113-0021, Japan.
[2] Mailing address: 5-36-2 Akazutsumi, Setagaya-ku, Tokyo 156-0044, Japan.
[3] We should write E_n, ξ_n, p_n etc. but we omit the suffix n hereafter.

CP1021, *Stochastic and Quantum Dynamics of Biomolecular Systems, Proceedings of the 5th Jagna International Workshop,* edited by C. C. Bernido and M. V. Carpio-Bernido

$$x > q + h \tag{5}$$

for example, with q and h defined by $V(q) = E$ and $h = q^\nu$ (with $0 < \nu < 1$ to be determined later. See Fig. 1 in [1]), $\Lambda^{(l)}$'s are given succesively by

$$\begin{aligned} \Lambda^{(0)}(\xi) &= Ae^{-k\xi}, \\ \Lambda^{(l)}(\xi) &= \frac{1}{-2k}\int_{\xi_h}^{\infty} e^{-k|\xi-\eta|} Q(\eta)\Lambda^{(l-1)}(\eta)\,d\eta, \qquad (l = 1, 2, \cdots), \end{aligned} \tag{6}$$

with

$$Q(\xi) = -\frac{5}{16p(x)^3}\left(\frac{dp(x)}{dx}\right)^2 + \frac{1}{4p(x)^2}\frac{d^2p(x)}{dx^2}, \tag{7}$$

and

$$\xi = \int_q^x p(x')^{1/2}dx' \quad \text{and} \quad \xi_h = \int_q^{q+h} p(x')^{1/2}dx'. \tag{8}$$

We assume the potential to have the form,

$$V(x) = a_{2m}x^{2m} + a_{2m-1}x^{2m-1} + \cdots + a_0 \quad (m = 1, 2, \cdots, a_{2m} > 0). \tag{9}$$

Then, $V(x)$ increases but Q decreases at $|x| \to \infty$.

The first term in the series (4) gives the well-known WKB approximate soluion to (1):

$$u^{(0)}(x) = \frac{1}{\left\{V(x)/E - 1\right\}^{1/4}} \exp\left[-\sqrt{\frac{E}{D}}\int_{\xi_h}^{x}\sqrt{\frac{V(x')}{E} - 1}\,dx'\right], \tag{10}$$

The present paper will solve the two problems:

(i) To prove the convergence of the series (4), which we call the iteration series.

It is well-known that the WKB approximation for $V(x)$ of the form of (9) with $m > 2$, if pushed to the higher orders, diverges to have the meaning only in the sense of asymptotic series (to which the Borel summability method applies). Then, our second problem is:

(ii) To explain how the solution to (1) can have two faces: convergent iteration series and the divergent WKB series.

ASYMPTOTIC SERIES

Before starting to solve the problems, let us explain what the asymptotic series is. Consider

$$F(x) = \int_0^\infty \frac{e^{-\eta}}{1+\eta x} d\eta \quad (x \geq 0) \tag{11}$$

for example. If one uses

$$\frac{1}{1+\eta x} = \sum_{n=0}^{N} (-\eta x)^n + \frac{(-\eta x)^{N+1}}{1+\eta x},$$

then term-by-term integration gives

$$F(x) = \sum_{n=0}^{N} (-1)^n n! x^n + R_N(x), \tag{12}$$

where

$$R_N(x) = (-1)^{N+1} x^{N+1} \int_0^\infty \frac{\eta^{N+1} e^{-\eta}}{1+\eta x} d\eta$$

satisfies

$$|R_N(x)| \leq x^{N+1} \int_0^\infty \eta^{N+1} e^{-\eta} d\eta = (N+1)! x^{N+1}. \tag{13}$$

Hence

$$|R_N(x)| = o(x^N) = o(N\text{th term of the series}) \quad \text{as} \quad x \to 0. \tag{14}$$

When this condition (14) holds, the series is said to be an asymptotic series of order N as $x \to 0$.

When $N \to \infty$, we write

$$F(x) \sim 1 - x + (2!)x^2 - (3!)x^3 + \cdots, \tag{15}$$

formally and say this series is divergent but asymptotic.

Borel Summability

If, for a divergent series,

$$F = a_0 + a_1 + a_2 + a_3 + \cdots, \tag{16}$$

the modified series

$$B(\eta) = a_0 + a_1 \frac{\eta}{1!} + a_2 \frac{\eta^2}{2!} + a_3 \frac{\eta^3}{3!} + \cdots \tag{17}$$

converges uniformly for $|\eta| < r$ for some r and

$$S_B = \int_0^\infty e^{-\eta} B(\eta) d\eta \tag{18}$$

converges, then (18) is said to be Borel sum of the series (16), and the series (16) Borel summable.

For the series (15), we have

$$B(\eta,x) = 1 - \eta x + (\eta x)^2 - (\eta x)^3 + \cdots = \frac{1}{1+\eta x} \qquad \text{for} \quad |\eta x| < 1, \tag{19}$$

so that the Borel sum,

$$S_{\mathrm{B}}(x) = \int_0^\infty \frac{1}{1+\eta x} e^{-\eta} d\eta, \tag{20}$$

reproduces $F(x)$ of (11).

CONVERGENCE OF THE ITERATION SERIES

Let us solve our Problem (i) posed at the end of INTRODUCTION. We restrict our attention to the region (5).

We consider large E and consequently q is large, so that the polynomial $V(x)$ is dominated by the highest order term $a_{2m}x^{2m}$ whose coefficient is assumed to be positive, $a_{2m} > 0$.

Crucial in the convergence proof is the estimate of p,

$$p(x) = \frac{V(x)}{E} - 1 = \frac{V(x)}{E}\left(1 - \frac{E}{V(x)}\right) \geq \frac{V(x)}{E}\left(1 - \frac{V(q)}{V(q+h)}\right)$$

valid because $V(x) \geq V(q+h)$ for (5), which gives

$$p(x) \geq \text{const.} \frac{V(x)}{E}\frac{h}{q}. \tag{21}$$

Consequently, from (7)

$$|Q(x)| \leq \text{const}\left\{\left(\frac{E}{V(x)}\frac{q}{h}\right)^3 \left(\frac{V(x)}{xE}\right)^2 + \left(\frac{E}{V(x)}\frac{q}{h}\right)^2 \frac{V(x)}{x^2 E}\right\},$$

so that, for large k,

$$|Q(x)| \leq \text{const.} \frac{E}{x^{2m+2}} k^{3(1-v)/m} \tag{22}$$

since $E = V(q) \sim aq^{2m}$ and $E = Dk^2$ gives $q \sim k^{1/m}$.

But,

$$\xi(x) = \int^x \sqrt{\frac{V(x')}{E} - 1}\, dx' \sim \sqrt{\frac{a}{E}} \frac{1}{m+1} x^{m+1}. \tag{23}$$

Hence, the estimate of Q:

$$|Q(\xi)| < \text{const.} \frac{k^{3(1-v)/m}}{\xi^2}. \tag{24}$$

From

$$\Lambda^{(l)}(\xi) = \frac{1}{-2k}\int_{\xi_h}^{\infty} e^{-k|\xi-\eta|}Q(\eta)\Lambda^{(l-1)}(\eta)\,d\eta,$$

we have

$$|\Lambda^{(l)}(\xi)| \leq \frac{1}{2k}\int_{\xi_h}^{\infty} e^{-k|\xi-\eta|}|Q(\eta)|\,e^{-k(\eta-\xi_h)}\left(\sup_{\eta'\geq\xi_h}\frac{|\Lambda^{(l-1)}(\eta')|}{e^{-k(\eta'-\xi_h)}}\right)d\eta, \qquad (25)$$

where existence of the sup will be shown later by (28). Then, using the estimate (24), the part of the integral I in (25) turns out to be:

$$I \leq \text{const.}k^{3(1-\nu)/m}\int_{\xi_h}^{\infty} e^{-k|\xi-\eta|}\frac{1}{\eta^2}e^{-k(\eta-\xi_h)}\,d\eta.$$

By integration by parts, we find

$$I \leq \text{const.}k^{3(1-\nu)/m}\frac{1}{\xi_h}e^{-k(\xi-\xi_h)}. \qquad (26)$$

But, by (21)

$$\begin{aligned}\xi_h &= \int_q^{q+h} p(x)^{1/2}dx > \frac{\text{const.}}{k^{(1-\nu)/2m}k}q^{m+1}\left\{\left(1+\frac{h}{q}\right)^{m+1}-1\right\} \\ &> \text{const.}\,k^{(3\nu-1)/2m}. \end{aligned} \qquad (27)$$

Hence, by (25), (26) and (27),

$$|\Lambda^{(l)}(\xi)| \leq \text{const.}\,k^{-\beta}e^{-k(\xi-\xi_h)}\sup_{\eta'\geq\xi_h}\frac{|\Lambda^{(l-1)}(\eta')|}{e^{-k(\eta'-\xi_h)}} \qquad (28)$$

justifying the sup estimate in (25), where

$$\beta = 1-\frac{3(1-\nu)}{m}+\frac{3\nu-1}{2m} = \frac{9\nu-7+2m}{2m}. \qquad (29)$$

Thus, using (28) repeatedly, we find

$$|\Lambda^{(l)}(\xi)| \leq (\text{const.}k^{-\beta})^l|A|e^{-k\xi_h}, \qquad (30)$$

so that, by choosing ν such that

$$\frac{7-2m}{9} < \nu < 1, \qquad (31)$$

we have the convergence, uniform in ξ, of the series (4) for sufficiently large k.

WKB TO HIGHER ORDER

We now proceed to solve problem (ii) posed at the end of INTRODUCTION.

Exponential form

We push the ordinary WKB approximation to the higher order. Put

$$\Lambda(\xi) = \exp\left[-k\xi + \int^{\xi} S_0(\eta)d\eta + \frac{1}{-2k}\int^{\xi} S_1(\eta)d\eta + \frac{1}{(-2k)^2}\int^{\xi} S_2(\eta)d\eta + \cdots\right], \tag{32}$$

where the lower limits of the integrals are arbitrary, and the first term in the exponent gives the ordinary WKB approximation (10). Then,

$$\begin{aligned}
\frac{d\Lambda}{d\xi} &= \left[-k + S_0(\xi) + \frac{1}{-2k}S_1(\xi) + \frac{1}{(-2k)^2}S_2(\xi) + \cdots\right]\Lambda(\xi), \\
\frac{d^2\Lambda}{d\xi^2} &= \left[\frac{dS_0(\xi)}{d\xi} + \frac{1}{-2k}\frac{dS_1(\xi)}{d\xi} + \cdots + \left(-k + S_0(\xi) + \frac{1}{-2k}S_1(\xi) + \cdots\right)^2\right]\Lambda(\xi).
\end{aligned}$$

The equation (3) gives:

$$\begin{aligned}
k^2 &: (-1)^2 = 1, \\
k^1 &: -2S_0 = 0, && S_0 = 0, \\
k^0 &: \frac{dS_0}{d\xi} + S_1 = Q && S_1 = Q, \\
(-2k)^{-1} &: \frac{dS_1}{d\xi} + S_2 + 2S_0S_1 = 0, && S_2 = -\frac{dS_1}{d\xi} = -\frac{dQ}{d\xi}, \\
(-2k)^{-2} &: \frac{dS_2}{d\xi} + S_3 + 2S_2S_0 + S_1^2 = 0, && S_3 = -\frac{dS_2}{d\xi} - S_1^2 = \frac{d^2Q}{d\xi^2} - Q^2, \\
(-2k)^{-3} &: \frac{dS_3}{d\xi} + S_4 + 2S_2S_1 = 0, && S_4 = -\frac{dS_3}{d\xi} - 2S_2S_1 = -\frac{d^3Q}{d\xi^3} + 2\frac{dQ^2}{d\xi}.
\end{aligned} \tag{33}$$

In general, we have $S_0 = 0$, $S_1 = Q$ and

$$S_n = -\frac{dS_{n-1}}{d\xi} - \sum_{j+l=n-1,\, j,l\geq 1} S_jS_l \quad (n = 2, 3, \cdots). \tag{34}$$

Thus, (34) solved step by step gives [2] [3]

$$\begin{aligned}
\Lambda(\xi) &= e^{-k\xi}\exp\left[\frac{1}{-2k}\int Q(\xi)d\xi - \frac{1}{(-2k)^2}\int \frac{dQ}{d\xi}d\xi + \frac{1}{(-2k)^3}\int\left(\frac{d^2Q}{d\xi^2} - Q^2\right)d\xi \cdots\right] \\
&= e^{-k\xi}\exp\left[\frac{1}{-2k}\int Q(\xi)d\xi - \frac{1}{(-2k)^2}Q(\xi) + \frac{1}{(-2k)^3}\left(\frac{dQ(\xi)}{d\xi} - \int^{\xi} Q(\xi)^2 d\xi\right)\right. \\
&\quad \left. + \frac{1}{(-2k)^4}\left(-\frac{d^2Q(\xi)}{d\xi^2} + 2Q(\xi)^2\right) + \cdots\right].
\end{aligned} \tag{35}$$

WKB series

If we expand the exponential in (32), we get

$$e^{k\xi}\Lambda(\xi) = L^{(0)}(\xi) + \frac{1}{-2k}L^{(1)}(\xi) + \frac{1}{(-2k)^2}L^{(2)}(\xi) + \cdots, \tag{36}$$

where

$$\begin{aligned}
L^{(0)} &= 1, \\
L^{(1)} &= \int Q d\xi, \\
L^{(2)} &= -Q + \frac{1}{2!}\left(\int Q d\xi\right)^2, \\
L^{(3)} &= \frac{dQ}{d\xi} - \int Q^2 d\xi - \frac{2}{2!}Q\int Q d\xi + \frac{1}{3!}\left(\int Q d\xi\right)^3, \\
L^{(4)} &= -\frac{d^2Q}{d\xi^2} + \frac{5}{2}Q^2 + \left(\int Q d\xi\right)\left(\frac{dQ}{d\xi} - \int Q^2 d\xi\right) \\
&\qquad -\frac{3}{3!}\left(\int Q d\xi\right)^2 Q + \frac{1}{4!}\left(\int Q d\xi\right)^4, \\
L^{(5)} &= \frac{d^3Q}{d\xi^3} - \frac{7}{2}\frac{dQ^2}{d\xi} + \int\left(\frac{dQ}{d\xi}\right)^2 d\xi - \frac{d^2Q}{d\xi^2}\int Q d\xi \\
&\quad +2\int Q^3 d\xi + \frac{5}{2}Q^2\int Q d\xi + Q\int Q^2 d\xi + \frac{1}{2}\frac{dQ}{d\xi}\left(\int Q d\xi\right)^2 \\
&\quad -\frac{1}{2}\left(\int Q d\xi\right)^2\int Q^2 d\xi - \frac{1}{6}Q\left(\int Q d\xi\right)^2 + \frac{1}{5!}\left(\int Q d\xi\right)^5. \\
&\vdots \qquad\qquad \vdots
\end{aligned} \tag{37}$$

ITERATION SERIES AND ASYMPTOTICS

We continue to consider the range (5), but from the point of view of iteration solution. Rewrite (6) in the form:

$$\Lambda(\xi) = e^{-k\xi} + \frac{1}{-2k}\int^{\xi} e^{-k(\xi-\eta)}Q(\eta)\Lambda(\eta)d\eta + \int_{\xi}^{\infty} e^{k(\xi-\eta)}Q(\eta)\Lambda(\eta)d\eta. \tag{38}$$

Or, put

$$\Lambda(\xi) = e^{-k\xi}M(\xi), \tag{39}$$

then

$$M(\xi) = 1 + \frac{1}{-2k}\left\{\int^{\xi} Q(\eta)M(\eta)d\eta + e^{2k\xi}\int_{\xi}^{\infty} e^{-2k\eta}Q(\eta)M(\eta)d\eta\right\}. \tag{40}$$

Now, the iteration gives

$$M(\xi) = M^{(0)}(\xi) + \frac{1}{-2k}M^{(1)}(\xi) + \frac{1}{(-2k)^2}M^{(2)}(\xi) + \cdots, \tag{41}$$

where

$$\begin{aligned} M^{(0)}(\xi) &= 1, \\ M^{(l)}(\xi) &= \int^{\xi} Q(\eta)M^{(l-1)}(\eta)d\eta + e^{2k\xi}\int_{\xi}^{\infty} e^{-2k\eta}Q(\eta)M^{(l-1)}(\eta)d\eta \quad (l = 1, 2, \cdots). \end{aligned} \tag{42}$$

Asymptotic expansion

For $l = 1$, we have

$$M^{(1)}(\xi) = \int^{\xi} Q(\eta)d\eta + e^{2k\xi}\int_{\xi}^{\infty} e^{-2k\eta}Q(\eta)d\eta. \tag{43}$$

If we look for an asymptotic expansion of Λ to the order $(-2k)^{-N}$, we can rewrite the second term on the right-hand side of this equation by repeating integration by parts $N+1$ times (N is arbitrary),

$$\begin{aligned} e^{2k\xi}\int_{\xi}^{\infty} e^{-2k\eta}Q(\eta)d\eta &= \sum_{n=1}^{N}\left(\frac{-1}{-2k}\right)^n \frac{d^{n-1}Q(\xi)}{d\xi^{n-1}} \\ &+ \left(\frac{-1}{-2k}\right)^{N+1}\left\{\frac{d^N Q(\xi)}{d\xi^N} + e^{2k\xi}\int_{\xi}^{\infty} e^{-2k\eta}\frac{d^{N+1}Q(\eta)}{d\eta^{N+1}}d\eta\right\}, \end{aligned}$$

which shows (α) that the repeated derivative causes $N!$ or the like to make the infinite series,

$$M^{(1)}(\xi) = \int^{\xi} Q(\eta)d\eta + \sum_{n=1}^{\infty}\left(\frac{-1}{-2k}\right)^n \frac{d^{n-1}Q(\xi)}{d\xi^{n-1}}, \tag{44}$$

diverge and (β) that the series is asymptotic to arbitrary order N.

Then, in the sense of asymptotic expansion, we get from (42)

$$M^{(2)}(\xi) = \int^{\xi} Q(\eta)M^{(1)}(\eta)d\eta + e^{2k\xi}\int_{\xi}^{\infty} e^{-2k\eta}Q(\eta)M^{(1)}(\eta)d\eta, \tag{45}$$

the second term of which can be transformed again by the repeated integration by parts:

$$\begin{aligned} e^{2k\xi}\int_{\xi}^{\infty} e^{-2k\eta}Q(\eta)M^{(1)}(\eta)d\eta &= \sum_{n=1}^{N}\left(\frac{-1}{-2k}\right)^n \frac{d^{n-1}Q(\xi)M^{(1)}(\xi)}{d\xi^{n-1}} \\ &+ \left(\frac{-1}{-2k}\right)^{N+1}\left\{\frac{d^N Q(\xi)M^{(1)}(\xi)}{d\xi^N} + e^{2k\xi}\int_{\xi}^{\infty} e^{-2k\eta}\frac{d^{N+1}Q(\eta)M^{(1)}(\eta)}{d\eta^{N+1}}d\eta\right\}. \end{aligned}$$

which shows again that the infinite series,

$$M^{(2)}(\xi) = \int QM^{(1)} d\xi + \sum_{n=1}^{\infty} \left(\frac{-1}{-2k}\right)^n \frac{d^{n-1} QM^{(1)}}{d\xi^{n-1}}, \tag{46}$$

is divergent but asymptotic. Substituting from (44),

$$\begin{aligned} M^{(2)}(\xi) = &\frac{1}{2!}\left(\int Q d\xi\right)^2 + \frac{1}{-2k}\left\{-\int^{\xi} Q^2 d\xi - Q\int Q d\xi\right\} \\ &+ \left(\frac{1}{-2k}\right)^2 \left\{\frac{5}{2}Q^2 + \frac{dQ}{d\xi}\int Q d\xi\right\} \\ &+ \left(\frac{1}{-2k}\right)^3 \left\{-\frac{7}{2}\frac{dQ^2}{d\xi} + \int \left(\frac{dQ}{d\xi}\right)^2 d\xi - \frac{d^2Q}{d\xi^2}\int Q d\xi\right\} + \cdots, \end{aligned} \tag{47}$$

where we have used

$$\int Q\frac{d^2Q}{d\xi^2} d\xi = \frac{1}{2}\frac{dQ^2}{d\xi} - \int \left(\frac{dQ}{d\xi}\right)^2 d\xi$$

and

$$\frac{1}{k!}\int Q(\xi)\left(\int Q d\xi\right)^k d\xi = \frac{1}{(k+1)!}\left(\int Q d\xi\right)^{k+1}. \tag{48}$$

From (42) similarly,

$$M^{(3)}(\xi) = \int QM^{(2)} d\xi + \sum_{n=1}^{\infty} \left(\frac{1}{2k}\right)^n \frac{d^{n-1} QM^{(2)}}{d\xi^{n-1}},$$

so that by (47),

$$\begin{aligned} M^{(3)}(\xi) = &\frac{1}{3!}\left(\int Q d\xi\right)^3 + \frac{1}{-2k}\left\{-\left(\int Q d\xi\right)\left(\int Q^2 d\xi\right) - \frac{1}{2}Q\left(\int Q d\xi\right)^2\right\} \\ &+ \left(\frac{1}{-2k}\right)^2 \left\{2\int Q^3 d\xi + \frac{5}{2}Q^2\int Q d\xi + Q\int Q^2 d\xi + \frac{1}{2}\frac{dQ}{d\xi}\left(\int Q d\xi\right)^2\right\} + \cdots, \end{aligned} \tag{49}$$

where we have used (48),

$$\int Q\left(\int Q^2 d\xi\right) d\xi + \int Q^2\left(\int Q d\xi\right) d\xi = \left(\int Q d\xi\right)\left(\int Q^2 d\xi\right),$$

and

$$\int \left(Q\frac{dQ}{d\xi}\int Q d\xi\right) d\xi = \frac{1}{2}Q^2\int Q d\xi - \frac{1}{2}\int Q^3 d\xi.$$

Further

$$M^{(4)}(\xi) = \frac{1}{4!}\left(\int Qd\xi\right)^4 + \frac{1}{-2k}\left\{-\frac{1}{2}\left(\int Qd\xi\right)^2 \int Q^2 d\xi - \frac{1}{6}Q\left(\int Qd\xi\right)^3\right\} + \cdots, \tag{50}$$

where we have used (48) and

$$\begin{aligned}\int\left(Q\int Qd\xi\int Q^2d\xi\right)d\xi &= \frac{1}{2}\left[\left\{\frac{d}{d\xi}\left(\int Qd\xi\right)^2\right\}\int Q^2d\xi\right]d\xi \\ &= \frac{1}{2}\left(\int Qd\xi\right)^2\int Q^2d\xi - \frac{1}{2}\int Q^2\left(\int Qd\xi\right)^2 d\xi,\end{aligned}$$

and

$$M^{(5)} = \frac{1}{5!}\left(\int Qd\xi\right)^5 + \cdots \tag{51}$$

by (48).

This calculation can be carried out to any order by the same algorithm, although we are unable to write down a general formula for $M^{(l)}$ at this moment.

COMPARISON OF WKB AND ITERATION SERIES

We have calculated Λ in two different ways, one by the WKB method and the other by asymptotic re-expansion of the iteration series.

Rearranging the terms in $M^{(l)}$ in ascending order of $(-2k)^{-n}$, or the order of $L^{(1)}, L^{(2)}, \cdots$, we get the horizontal rows in the following table.

Table 1. Comparison of the WKB and the Iteration Series

$M^{(1)}$	$\int Qd\xi$	$-Q$	$\frac{dQ}{d\xi}$
$M^{(2)}$		$\frac{1}{2!}\left(\int Qd\xi\right)^2$	$-\int Q^2d\xi - Q\int Qd\xi$
$M^{(3)}$			$\frac{1}{3!}\left(\int Qd\xi\right)^3$
Sum	$L^{(1)}$	$L^{(2)}$	$L^{(3)}$

Table 1, cont'd.

$M^{(1)}$	$-\dfrac{d^2Q}{d\xi^2}$	$\dfrac{d^3Q}{d\xi^3}$
$M^{(2)}$	$\dfrac{5}{2}Q^2+\dfrac{dQ}{d\xi}\int Qd\xi$	$-\dfrac{7}{2}\dfrac{dQ^2}{d\xi}+\int\left(\dfrac{dQ}{d\xi}\right)^2 d\xi-\dfrac{d^2Q}{d\xi^2}\int Qd\xi$
$M^{(3)}$	$-\left(\int Qd\xi\right)\int Q^2d\xi$	$+2\int Q^3d\xi+\dfrac{5}{2}Q^2\int Qd\xi$
	$-\dfrac{1}{2}Q\left(\int Qd\xi\right)^2$	$+Q\int Q^2d\xi+\dfrac{1}{2}\dfrac{dQ}{d\xi}\left(\int Qd\xi\right)^2$
$M^{(4)}$	$\dfrac{1}{4!}\left(\int Qd\xi\right)^4$	$-\dfrac{1}{2}\left(\int Qd\xi\right)^2\int Q^2d\xi-\dfrac{1}{6}Q\left(\int Qd\xi\right)^3$
$M^{(5)}$		$+\dfrac{1}{5!}\left(\int Qd\xi\right)^5$
Sum	$L^{(4)}$	$L^{(5)}$

In this Table, the vertical column-by-column sums all agree with the corresponding $L^{(j)}$ in (37), showing that the WKB series agree with the asymptototic re-expansion of the convergent iteration series to order $(-2k)^{-5}$. We are trying to establish that these two agree to all orders.

CONCLUSION

Eigenvalue problem for the Schrödinger equation in one dimension for a polynomial potential with the highest order term having an even power and a positive coefficient, is solved for high eigenvalues, $E_n = Dk_n^2$ in two different ways after Liouville transformation, which keeps the form of the Schrödinger equation invariant but changes the potential into the one decreasing at infinity:

(a) Converting the differential equation into integral equation (**??**) and solving it iteratively,

(b) By the WKB method, expanding the logarithm of the eigenfunction in a power series in $1/k_n$.

While the series solution from (b) is known to diverge, we have shown that the solution from (a) converges for sufficiently high k_n.

We have shown that the convergent iterative solution from (a), if re-expanded in powers of $1/k_n$, gives a series, divergent but asymptotic, which agrees with the series solution from (b). Actually, we have been able to show the agreement only up to $O(k_n^{-5})$, but we believe and are now trying to establish that it holds to all orders.

If this is true, then the divergent WKB series can be reorganized to be a convergent series, which can in fact be obtained by iterative procedure (a).

ACKNOWLEDGMENTS

It is our great pleasure to dedicate this paper to Professors Frederik Wiegel and Ludwig Streit on the occasion of their 70th birthday. We have long been benefited greatly from discussions with them.

REFERENCES

1. T. Nakamura, H. Ezawa and K. Watanabe, Time-Development of the Explosion of a Stochastic Process having a Repulsive Drift of Polynomial Growth, This Proceeding.
2. K. Terasawa ed., *Mathematical Methods for Scientists - 2. Applications*, Iwanami (1960).
3. Imai obtained the formula up to $O(k^{-4})$. I. Imai, Phys. Rev. **74**, 113 (1948); **80** 1112 (1950).

Charged Particle in a Time-dependent Electric Field: A White Noise Functional Approach

E.B. Gravador, J.B. Bornales and M.J. Liwanag

Physics Department, Mindanao State University - Iligan Institute of Technology, Iligan City 9200 Philippines

Abstract. The propagator for a charged particle in a time-dependent electric field is calculated following Hida and Streit's framework [1] where the propagator is the T-transform of Feynman functional. However, we have to regard the potential $V \quad -qE(\tau)x \equiv \sqrt{\frac{m}{\hbar}}\dot{\xi}x$ following C.C. Bernido and M.V. Carpio-Bernido's [2] prescription of time-dependent potentials. The result agrees with the limiting form of Eqn(16) of N.Morgenstern Horing, H.L.Cui, and G. Fiorenza [3], when the magnetic field is switched off , and Eqn(17) of [3] when the electric field is constant in time.

Keywords: white noise functional, Feynman functional, T-transform, quantum mechanical propagator
PACS: 02.50.Ey, 03.65.Db, 05.40.-a

INTRODUCTION

In 1983, Hida and Streit [1] came up with an expression for the quantum mechanical propagator as the T-transform of Feynman functional. This is the well-known Eqn(2.20) which has been applied to a potential which depends at most quadratically in the White Noise variable ω. In particular, the propagator for the harmonic oscillator has been obtained in closed form using Eqn(2.20) [1].

In this paper, we are interested in applying Eqn(2.20) to a potential which is linear in ω. A charged particle in a time-dependent electric field has the potential $V = -qE(\tau)x$ which is linear in ω upon parametrization of x in terms of Brownian fluctuations from a sure path. Apparently, Eqn(2.20) with the choice of the test function $\xi(\tau) = -\sqrt{\frac{\hbar}{m}}\, qE(\tau)$ does not lead directly to known results (see for instance Horing, Cui and Fiorenza [3]). However, by treating the potential $V = -qE(\tau)x$ as $V = \sqrt{\frac{m}{\hbar}}\dot{\xi}(t)x$ according to the formalism for time-dependent potentials given by C.C. Bernido and M.V. Carpio-Bernido [2] we are able to obtain the limiting form of Eqn(16) of Horing,Cui and Fiorenza [3], when the magnetic field is switched off , and Eqn(17) of [3] when the electric field is constant in time.

In the next section, we review Hida and Streit's framework leading to Eqn(2.20) and mention C.C. Bernido and M.V. Carpio-Bernido's treatment of time-dependent potentials. We then apply these techniques in obtaining the propagator for a charged particle in a time-dependent electric field.

CP1021, *Stochastic and Quantum Dynamics of Biomolecular Systems, Proceedings of the 5th Jagna International Workshop,* edited by C. C. Bernido and M. V. Carpio-Bernido

THE PROPAGATOR AS T-TRANSFORM OF FEYNMAN FUNCTIONAL $I(\omega)$

In general, one can write the propagator as the T-transform or in terms of the S-transform of Feynman functional

$$K(y_1,y_2;0,t) = (TI)(\xi) = C(\xi)(SI)(i\xi) \quad . \tag{1}$$

For a Brownian functional of the form

$$\Phi(\omega) = \mathcal{N} \exp\left(-\frac{1}{2}\langle\omega,(K+L)\omega\rangle + i\langle\omega,g\rangle\right)\delta(B(t)-y) \tag{2}$$

the S-transform is given by

$$\begin{aligned}(SI)(\xi) &= \frac{1}{2\pi}C(\xi)\mathcal{N}\int e^{-i\lambda y}d\lambda\int e^{-\frac{1}{2}\langle\omega,(K+L)\omega\rangle} \\ &\times\exp\left[\langle\omega,\xi\rangle + i\langle\omega,g\rangle + i\lambda\left\langle\omega,\chi_{[0,t]}\right\rangle\right]d\mu(\omega) \quad ,\end{aligned} \tag{3}$$

and the corresponding T-transform is

$$\begin{aligned}(T\Phi)(\xi) = &\left[2\pi t\left\langle e,N^{-1}e\right\rangle det(1+L(1+K)^{-1})\right]^{-1/2} \\ &\times\exp\left[-\frac{1}{2}\left\langle(\xi+g),N^{-1}(\xi+g)\right\rangle\right] \\ &\times\exp\left[-\frac{1}{2\left\langle e,N^{-1}e\right\rangle}\left(\frac{y}{\sqrt{t}}-i\left\langle(\xi+g),N^{-1}e\right\rangle\right)\right. \\ &\left.\times\left(\frac{y}{\sqrt{t}}-i\left\langle e,N^{-1}(\xi+g)\right\rangle\right)\right] \quad .\end{aligned} \tag{4}$$

This is Eqn(2.20) of [1].

C.C. Bernido and M.V. Carpio-Bernido [2] consider time-dependent potentials $V_2 = \dot{\xi}(\tau)x$ and velocity-dependent potentials $V_1 = -c\,\dot{x}\xi(\tau)$ where ξ is the test function which can be determined from the potential. In the next section, we construct the Feynman functional and the propagator for a charged particle in a time-dependent electric field.

THE FEYNMAN FUNCTIONAL FOR A CHARGED PARTICLE IN A TIME-DEPENDENT ELECTRIC FIELD

The Lagrangian of the system being considered is $\mathcal{L}(x,\dot{x}) = \frac{1}{2}m\dot{x}^2 + qE(\tau)x$, where E is the intensity of the electric field and $(-q)$ is the charge of the particle. The Feynman functional is

$$I = \exp\left[\frac{i}{\hbar}\mathcal{S}(X)\right] = \exp\left[\frac{i}{\hbar}\int\left(\frac{1}{2}m\dot{x}^2 + qE(\tau)x\right)d\tau\right] \quad . \tag{5}$$

To implement a White Noise Functional approach, the paths are parametrized as Brownian fluctuations from the sure path, $x(\tau) = y(\tau) + \sqrt{\frac{\hbar}{m}}B(\tau)$, $0 < \tau < t$. $\omega = \dot{B}$ is interpreted as a generalized function in an infinite dimensional space. The set $\{\dot{B}(\tau);\ \tau\, \varepsilon\, R\}$ of generalized functions constitute a continuum coordinate system.

Hida and Streit [1] regard the Feynman integral as the expectation of White Noise functional. The propagator is evaluated as the T-transform of the Feynman functional. The propagator for the charged particle in a time-dependent electric field as the expectation of Feynman functional is then given by

$$\begin{aligned} K(y_1,y_2;0,t) \quad &= \mathbf{E}\Big(\mathcal{N} \exp\left[\tfrac{(i+1)}{2}\int_0^t \omega^2\, d\tau\right] \exp\left[i\sqrt{\tfrac{m}{\hbar}}\int_0^t \dot{y}\,\omega\, d\tau\right] \\ &\times\ \sqrt{\tfrac{m}{\hbar}}\,\delta\left(B(t) - \sqrt{\tfrac{m}{\hbar}}(y_2 - y(t))\right) \exp\left[\tfrac{im}{2\hbar}\int_0^t \dot{y}^2 d\tau\right] \\ &\times\ \exp\left[\tfrac{i}{\hbar}\int_0^t qE(\tau)y(\tau)d\tau + \frac{i}{\hbar}\int_0^t \left\langle \omega, \sqrt{\frac{\hbar}{m}}\, qE(\tau) \right\rangle d\tau\right]\Big)\ . \end{aligned} \tag{6}$$

Using Eqn(2.20) of [1] or Eqn(4), we can identify from Eqn(6) the following: $K = -(1+i)\chi^2_{[0,t]}, L = 0\,, g = \sqrt{\frac{m}{\hbar}}\dot{y}$, the test function $\xi(\tau) = -\sqrt{\frac{\hbar}{m}}\, qE(\tau)$ and $y = \sqrt{\frac{m}{\hbar}}(y_2 - y(t))$. We find that $N^{-1} = \frac{i}{\chi^2_{[0,t]}}$ and $\langle e, N^{-1}e\rangle = i$.

The choice of the test function $\xi(\tau) = -\sqrt{\frac{\hbar}{m}}\, qE(\tau)$, however, does not lead us to known results (see for instance, N.Morgenstern Horing, H.L.Cui, and G. Fiorenza (1986)[3]) . C.C.Bernido and M.V. Carpio-Bernido [2] suggest that we focus on the object $\langle \omega, \dot{\xi}\rangle$, instead. Let us define time-dependent potential $V_2 = \sqrt{\frac{m}{\hbar}}\dot{\xi}x = -qE(\tau)x$ and denote the corresponding propagator as $K_{V_2}(y_1,y_2;0,t)$. Let us also define velocity-dependent potential $V_1 = -c\xi(\tau)\dot{x} = -\int_0^\tau qE(s)\, ds\, \dot{x}(\tau)$ and denote the corresponding propagator as $K_{V_1}(y_1,y_2;0,t)$. The two propagators are related as $K_{V_2}(y_1,y_2;0,t) = K_{V_1}(y_1,y_2;0,t)\exp\left(i\sqrt{\frac{m}{\hbar}}\xi(t)y_2 - i\sqrt{\frac{m}{\hbar}}\xi(0)y_1\right)$ (see Eqn(50) of [2]). We have the prescription then that we solve for $K_{V_1}(y_1,y_2;0,t)$ first and then obtain the propagator for a charged particle in time-dependent electric field. Since it is difficult to obtain the propagator for a charged particle in time-dependent electric field directly, so we make a time-transformation from $V_2 = \sqrt{\frac{m}{\hbar}}\dot{\xi}x$ to $V_1 = -c\xi(\tau)\dot{x}$. We work in the transform space and calculate for $K_{V_1}(y_1,y_2;0,t)$. Finally, we obtain $K_{V_2}(y_1,y_2;0,t)$ as an inverse-transformation.

Going back to Eqn(2.20) of [1] with $\xi(\tau) = -\sqrt{\frac{\hbar}{m}}\int_0^\tau qE(s)\, ds$, we write

$$\begin{aligned} K_{V_1}(y_1,y_2;0,t) = \left(\tfrac{m}{2\pi i\hbar t}\right)^{\frac{1}{2}} \exp\Big[-\tfrac{1}{2}\Big\langle\Big(-\sqrt{\tfrac{\hbar}{m}}\int_0^\tau qE(s_1)ds_1 + \sqrt{\frac{m}{\hbar}}\dot{y}\Big)\,, \\ \times\ \frac{i}{\chi^2_{[0,t]}}\Big(-\sqrt{\tfrac{\hbar}{m}}\int_0^\tau qE(s_2)ds_2 + \sqrt{\frac{m}{\hbar}}\dot{y}\Big)\Big\rangle\Big] \end{aligned}$$

$$\times \exp\left[-\tfrac{1}{2i}\left(\tfrac{y}{\sqrt{t}} - i\left\langle\left(-\sqrt{\tfrac{\hbar}{m}}\int_0^\tau qE(s_1)ds_1 + \sqrt{\frac{m}{\hbar}}\dot{y}\right), \tfrac{i}{\chi^2_{[0,t]}}\tfrac{\chi_{[0,t]}}{\sqrt{t}}\right\rangle\right)\right.$$

$$\left.\times \left(\tfrac{y}{\sqrt{t}} - i\left\langle\tfrac{\chi_{[0,t]}}{\sqrt{t}}, i\left(-\sqrt{\tfrac{\hbar}{m}}\int_0^\tau qE(s_2)ds_2 + \sqrt{\frac{m}{\hbar}}\dot{y}\right)\right\rangle\right)\right]$$

$$\times \exp\left[\tfrac{im}{2\hbar}\int_0^t \dot{y}^2 d\tau\right] \quad . \tag{7}$$

We evaluate the integrals in Eqn(7) and use the result to obtain the required propagator. We find

$$K_{V_2}(y_1,y_2;0,t) = \left(\tfrac{m}{2\pi i\hbar t}\right)^{\frac{1}{2}} \exp\left[\tfrac{im}{2\hbar t}(y_2-y_1)^2 - \tfrac{iq^2\hbar}{2m}\int_0^t d\tau \int_0^\tau E(s_1)\,ds_1 \int_0^\tau E(s_2)\,ds_2\right]$$

$$\times \exp\left[-\tfrac{iq}{t}(y_2-y_1)\int_0^t d\tau \int_0^\tau E(s_1)\,ds_1 + \frac{iq^2\hbar}{2mt}\left(\int_0^t d\tau \int_0^\tau E(s_1)\,ds_1\right)^2\right]$$

$$\times \exp\left[iq(y_2-y_1)\int_0^t E(\tau)\,d\tau\right] \exp\left[-iqy_2\int_0^t E(\tau)\,d\tau\right]. \tag{8}$$

This is the limiting form of Eqn(16) of [3] when the magnetic field is switched off. As a check, Eqn(8) reduces to Eqn(17) of [3] when the electric field is constant in time .

ACKNOWLEDGMENTS

We wish to thank Christopher C. Bernido and Maria Victoria Carpio-Bernido for their encouragement, nurture and support. We also acknowledge the help of the Physics Department of MSU-IIT in making possible our attendance in the workshop.

REFERENCES

1. L. Streit, and T. Hida , "Generalized Brownian functionals and the Feynman path integral," *Stochastic Processes and their Applications* , **16**, 55–69 (1983).
2. C.C. Bernido, and M.V. Carpio-Bernido, "White Noise Analysis and the Feynman Path Integral" in *Proceedings of the 3rd Jagna International Workshop: Functional Integrals in Stochastic and Quantum Dynamics.*, edited by C.C. Bernido, M.V. Carpio-Bernido, and L. Streit , Central Visayan Institute Foundation, Jagna, Bohol, 2001, pp. 18–31 .
3. N.Morgenstern Horing, H.L.Cui, and G. Fiorenza, "Nonrelativistic Schrodinger Green's function for crossed time-dependent electric and magnetic fields," *Physical Review A* **34**, 612-615 (1986).
4. D. de Falco, and D.C. Khandekar, "Applications of white noise calculus to the computation of Feynman integrals," *Stochastic Processes and Applications* **29** 257-266 (1988).
5. M. Grothaus, D.C Khandekar, J.L. da Silva, and L. Streit, "The Feynman integral for time-dependent anharmonic oscillators," *Journal of Mathematical Physics* **38** 3278–3299 (1997).

White Noise Path Integral Treatment of a Two-dimensional Dirac Oscillator in a Uniform Magnetic Field

Lyndon D. Bastatas and Jinky B. Bornales

Physics Department, MSU - Iligan Institute of Technology
Andres Bonifacio Ave., Tibanga, Iligan City, 9200 Philippines

Abstract. White noise path integral prescription is applied to solve the Dirac equation for a two-dimensional Dirac oscillator in a uniform magnetic field. The energy spectrum obtained agrees with the result obtained by Villalba and Maggiolo [1] using the differential approach.

Keywords: relativistic quantum mechanics, functional analysis, Brownian motion
PACS: 31.30.J, 03.65.Db, 05.40.Jc

INTRODUCTION

The Dirac oscillator is the relativistic analog of harmonic oscillator which was introduced by Ito, et al. [2] and later developed by Monshinsky, et al. [3]. It was conceived by introducing a correction to the momenta of the free Dirac particle with a linearly growing potential that is proportional to the particle's oscillation frequency. In Galilean limit, the Dirac oscillator behaves like a harmonic oscillator with a strong spin-orbit coupling.

Several authors [1, 4, 5, 6] have explored the system and it is only recently that the path integral approach was used to solve the Dirac oscillator [7, 8]. In [7, 8], the path integrals are evaluated using the Grassmann-number formulation. It is the aim of this paper to solve the path integral of the two-dimensional Dirac oscillator in a uniform magnetic field in the context of white noise analysis. In Section II, we illustrate how the path integral set-up for the system is achieved by considering the iterated Dirac equation as performed in [9, 10, 11]. In Section III, we demonstrate how the path integral is evaluated in the framework of white noise analysis following [13]. We then solve the Green function of the system and finally obtain its energy spectrum. All throughout in our calculations, we employ the natural units, i.e., $\hbar = c = 1$.

PATH INTEGRAL FOR THE DIRAC OSCILLATOR IN A UNIFORM MAGNETIC FIELD

The Green function $G(\mathbf{r}', \mathbf{r}'')$ for a spin $\frac{1}{2}$ particle which satisfies the Dirac equation is given by

$$(M - \hat{D})G(\mathbf{r}'', \mathbf{r}') = \delta(\mathbf{r}'' - \mathbf{r}') \tag{1}$$

CP1021, *Stochastic and Quantum Dynamics of Biomolecular Systems, Proceedings of the 5th Jagna International Workshop,* edited by C. C. Bernido and M. V. Carpio-Bernido

with M as the mass of the charged particle. For a two-dimensional Dirac oscillator in the presence of a constant magnetic field, the operator $\hat{D} = -\beta\alpha \cdot (\mathbf{p} - iMf\beta\mathbf{r} - e\mathbf{A}) + \beta E$ where β and $\alpha = \alpha\hat{i} + \alpha\hat{j}$ are Dirac matrices, $\mathbf{p} = p_x\hat{i} + p_y\hat{j}$ is the momentum operator, f as the particle's oscillation frequency, $\mathbf{r} = x\hat{i} + y\hat{j}$ is the position vector, E the energy and $\mathbf{A} = -\frac{yB}{2}\hat{i} + \frac{xB}{2}\hat{j}$ is the vector potential.

Equation (1) can be iterated by expressing the Green function, $G(\mathbf{r}'',\mathbf{r}')$, as

$$G(\mathbf{r}'',\mathbf{r}') = (M + \hat{D})g(\mathbf{r}'',\mathbf{r}') \tag{2}$$

where $g(\mathbf{r}'',\mathbf{r}')$ is the Green function satisfying the iterated Dirac equation $(M^2 - \hat{D}^2)g(\mathbf{r}'',\mathbf{r}') = \delta(\mathbf{r}'' - \mathbf{r}')$. The Green function, $g(\mathbf{r}'',\mathbf{r}')$, can be written as [9, 10, 11]

$$g(\mathbf{r}'',\mathbf{r}') = \frac{i}{2M}\int_0^{\infty} \langle \mathbf{r}''| \exp\{-i(H_{eff}\Lambda)\} |\mathbf{r}'\rangle \, d\Lambda \tag{3}$$

where $H_{eff} = \frac{M^2 - \hat{D}^2}{2M}$ is the effective Hamiltonian and $|\mathbf{r}\rangle = |x,y\rangle\,|\chi\rangle$ where $|\chi\rangle$ is a two-component spinor.

Since we are dealing with two-component spinor, we adopt the following representation in terms of Pauli matrices: $\beta = \sigma_z$; $\beta\alpha_x = i\sigma_x$; and $\beta\alpha_y = i\sigma_y$. Squaring the operator $\hat{D}$ and using the properties of Pauli matrices, the effective Hamiltonian, H_{eff}, becomes $H_{eff} = \frac{p_x^2 + p_y^2}{2M} + \frac{M\bar{f}}{2}(x^2 + y^2) - \bar{f}(xp_y - yp_x) - 2S_z\bar{f} - \frac{E^2 - M^2}{2M}$ with $S_z = \frac{1}{2}\sigma_z$ and $\bar{f} = f + \frac{eB}{2M}$. Here, $[H_{eff}, S_z] = 0$ and the eigenvalue of S_z once acting on an eigenspinor is $m_s = \pm\frac{1}{2}$.

Making use of the completeness relation $\sum_s |\xi_s >< \xi_s| = 1$, we can express Eq.(3) as

$$g(\mathbf{r}'',\mathbf{r}') = \frac{i}{2M}\sum_s \int_0^{\infty} \eta_s\eta_s^* \langle x''y''| \exp\{-i(H_s\Lambda)\} |x'y'\rangle \, d\Lambda \tag{4}$$

with $\eta_s =< \chi|\xi_s >$, $\eta_s^* =< \xi_s|\chi >$ and $H_s = \frac{p_x^2 + p_y^2}{2M} + \frac{M\bar{f}}{2}(x^2 + y^2) - \bar{f}(xp_y - yp_x) - 2m_s\bar{f} - \frac{E^2 - M^2}{2M}$. The integrand in Eq.(4) can be visualized as a propagator of the Dirac oscillator in a uniform magnetic field propagating in a time-like parameter Λ and can be expressed as a sum of over-all-paths [12], i.e.,

$$\langle x''y''| \exp\{-iH_s\Lambda\} |x'y'\rangle = \int \exp\{iS\} D^{\infty}[xy] \tag{5}$$

where S is the effective action of the system given by

$$S = \left(2m_s\bar{f} + \frac{E^2 - M^2}{2M}\right)\Lambda + \int_0^{\Lambda} \left\{\frac{M}{2}(\dot{x}^2 + \dot{y}^2) + M\bar{f}(x\dot{y} - y\dot{x})\right\} d\lambda. \tag{6}$$

With Eq.(6), the integrand of Eq.(4) can be written as $\langle x''y''| \exp\{-i(H_s\Lambda)\} |x'y'\rangle = \exp\left[i\Lambda\left(2m_s\bar{f} + \frac{E^2 - M^2}{2M}\right)\right] K(x'',y'';x',y')$ where

$$K(x'',y'';x',y') = \int \exp\left\{\int_0^{\Lambda} \left[\frac{iM}{2}(\dot{x}^2 + \dot{y}^2) + iM\bar{f}(x\dot{y} - y\dot{x})\right] d\lambda\right\} D[xy]. \tag{7}$$

We shall evaluate Eq.(7) using the white noise functional approach. This will be done in the next section.

WHITE NOISE PATH INTEGRATION FOR A DIRAC OSCILLATOR

To cast Eq.(7) in the language of white noise analysis, the paths of the particle are first parametrized in terms of the white noise variable ω, i.e., $x(\lambda)=x'+\sqrt{\frac{1}{M}}\int_0^\Lambda \omega_x(\lambda)d\lambda$ and $y(\lambda)=y'+\sqrt{\frac{1}{M}}\int_0^\Lambda \omega_y(\lambda)d\lambda$ where $\omega_x(\lambda)=dB_x(\lambda)/d\lambda$ and $\omega_y(\lambda)=dB_y(\lambda)/d\lambda$. Here, $B_x(\lambda)$ and $B_y(\lambda)$ are the Brownian motions along x and y coordinates, respectively. With the path parametrizations, the exponential term in the integrand of Eq.(7) then contains an expression similar to the Levy's stochastic area where the evaluation of which can be done in the probability space of a one-dimensional white noise space [13], that is

$$\left.\begin{array}{rclcrcl} B_x(\lambda) &=& \int_0^\lambda \omega(\lambda)d\lambda &;& dB_x(\lambda) &=& \omega(\lambda)d\lambda, \\ B_y(\lambda) &=& \int_{-\lambda}^0 \omega(\lambda)d\lambda &;& dB_x(\lambda) &=& \omega(-\lambda)d\lambda. \end{array}\right\} \tag{8}$$

The final endpoints, x'' and y'' are pinned down through the used of the Donsker delta functions $\delta(x'-x''+\sqrt{\frac{1}{M}}B_x(\lambda))$ and $\delta(y'-y''+\sqrt{\frac{1}{M}}B_y(\lambda))$. With the path parametrizations and the application of Eq.(8), together with the Fourier representation of the Donsker delta functions, Eq.(7) can then be expressed as

$$K(x'',y'';x',y')=\frac{1}{(2\pi)^2}\int dp_x dp_y \exp[ip_x(x'-x'')]\exp[ip_y(y'-y'')]$$
$$\times\int N_{xy}\exp(i\langle\omega,\xi\rangle)\exp\left(-\frac{1}{2}\langle\omega,K\omega\rangle\right)\exp\left(-\frac{1}{2}\langle\omega,L\omega\rangle\right)d\mu(\omega) \tag{9}$$

where $\xi=\frac{1}{\sqrt{M}}(p_x-\bar{f}y')\chi_{[0,\Lambda]}+\frac{1}{\sqrt{M}}(p_y+\bar{f}x')\chi_{[-\Lambda,0]}$, $K=-(i+1)$ and $L=-4i\bar{f}F_{S_T}(\lambda_1,\lambda_2)$ with $F_{S_T}=-\frac{1}{4}\left[\chi_{[0,\Lambda]}(\lambda_1)\chi_{[-\lambda_1,0]}(\lambda_2)+\chi_{[0,\Lambda]}(\lambda_2)\chi_{[-\lambda_2,0]}(\lambda_1)\right]+\frac{1}{4}\left[\chi_{[-\Lambda,0]}(\lambda_1)\chi_{[0,-\lambda_1]}(\lambda_2)+\chi_{[-\Lambda,0]}(\lambda_2)\chi_{[0,\ \lambda_2]}(\lambda_1)\right]$. The term $\chi_{[a,b]}(\lambda_k)$, $k=1,2$, with $[a,b]$ as the limit of integration, denotes the integration over λ_k.

We note that the integration over $d\mu(\omega)$ in Eq.(9) is just the T-transform of the functional $\Phi=N_{xy}\exp(i\langle\omega,\xi\rangle)\exp\left(-\frac{1}{2}\langle\omega,K\omega\rangle\right)\exp\left(-\frac{1}{2}\langle\omega,L\omega\rangle\right)$ in the white noise analysis. The calculations of the T-transform of this type of functional has already been done by D. de Falco and D.C. Khandekar [13]. Using the results obtained in [13] and integrating Eq.(9) over dp_x and dp_y yields

$$\begin{aligned} K(x'',y'';x',y') &= \frac{2M\bar{f}}{\pi}\exp\left[iM\bar{f}(x'y''-y'x'')\right]\exp\left\{-\frac{M\bar{f}}{2}\left((x''-x')^2+(y''-y')^2\right)\right\} \\ &\times \sum_{n=0}^{\infty}\sum_{m=0}^{\infty}\frac{(-1)^{-n}2^{-2n}}{m!(n-m)!}H_m(\sqrt{M\bar{f}}(x''-x'))H_{2(n-m)}(\sqrt{M\bar{f}}(y''-y')) \\ &\times \exp\left[-i\Lambda\bar{f}(2n+1)\right] \end{aligned} \tag{10}$$

where we have expressed the results in terms of the Hermite polynomials. Substituting Eq.(10) to Eq.(4) and then integrating over Λ, we obtain the second order Green function

$$\begin{aligned} g(\mathbf{r}'',\mathbf{r}') &= \frac{\bar{f}}{\pi}\sum_s \eta_s\eta_s^* \exp\left[iM\bar{f}(x'y''-y'x'')\right]\exp\left\{-\frac{M\bar{f}}{2}\left((x''-x')^2+(y''-y')^2\right)\right\} \\ &\times \sum_{n=0}^{\infty}\sum_{m=0}^{\infty}\frac{(-1)^{-n}2^{-2n}}{m!(n-m)!}H_m(\sqrt{M\bar{f}}(x''-x'))H_{2(n-m)}(\sqrt{M\bar{f}}(y''-y')) \\ &\times \left\{-2m_s\bar{f}-\frac{E^2-M^2}{2M}+\bar{f}(2n+1)+i\kappa\right\}^{-1}, \end{aligned} \tag{11}$$

in the limit $\kappa \to 0$. With Eq.(11), the Green function, $G(\mathbf{r}'',\mathbf{r}')$, for the first order Dirac equation can be readily obtain through Eq.(2).

We can obtain the energy spectrum from the poles of the Green function $g(\mathbf{r}'',\mathbf{r}')$ in Eq.(11). Shifting $n \to n-\frac{1}{2}|k+1|+\frac{1}{2}(k+\varepsilon)$ where $k \leq n-1$ is an integer and $\varepsilon = \pm 1$, for positive energy states (when $m_s = +\frac{1}{2}$, $\varepsilon = +1$ and $k > 0$), we have $E^2-M^2 = 4nM\bar{f} = 4n\left(Mf+\frac{eB}{2}\right)$ and for negative energy states (when $m_s = -\frac{1}{2}$, $\varepsilon = -1$ and $k < 0$), $E^2-M^2 = 4M\bar{f}\left(n+\frac{1}{2}-k'\right) = 4\left(Mf+\frac{eB}{2}\right)\left(n+\frac{1}{2}-k'\right)$ where $k' = k+\frac{1}{2}$. These results agree with the energy spectrum found in [1].

ACKNOWLEDGMENTS

The authors are grateful to Dr. C.C. Bernido and Dr. M.V.C. Bernido for their very enlightening lectures on “white noise and stochastic processes” during their several visits in MSU-IIT and for the opportunity given to the authors to participate to the 5th Jagna International Workshop.

REFERENCES

1. V. M. Villalba, and A. R. Maggiolo, *Eur. Phys. J.B.* **22**, 31-35 (2001).
2. D. Ito, K. Mori, and Carreri, *Nuovo Cim. A* **51**, 1119 (1967).
3. M. Moshinsky, and A. Szczepaniak, *J. Phys. A* **22**, 817-819 (1989).
4. J. Benitez, R. Martinez-y-Romero, H. N. Nunez-Yepez, and A. L. Brito, *Phys. Rev. Lett.* **64**, 1643-1645 (1990).
5. R. P. Martinez-y-Romero, H. N. Nunez-Yepez, and A. L. Salas-Britos, *Eur. J. Phys.* **16**, 135-141 (1995).
6. V. M. Villalba, *Phys. Rev. A* **49**, 586-587 (1994).
7. R. Rekioua, and T. Boudjedaa, *Eur. Phys. J.C* **49**, 1091-1098 (2007).
8. H. Haouat, and L. Chetouani, *Z. Naturforsch.* **62**A, 34-40 (2007).
9. M. A. Kayed, and A. Inomata, *Phys. Rev. Lett.* **53**, 107-110 (1984).
10. C. C. Bernido, M. V. Bernido, and N. S. Lim, *Phys. Lett. A* **231**, 395-401 (1997).
11. J. Bornales, C. C. Bernido, and M. V. Carpio-Bernido, *Phys. Lett. A* **260**, 447-452 (1999).
12. R. P. Feynman, and A. R. Hibbs, *Quantum Mechanics and Path Integrals*, McGraw-Hill, NewYork, 1965.
13. D. de Falco and D.C. Khandekar, *Stoch. Proc. and Appl.* **29**, 257-266, (1988).
14. L. Streit and T. Hida, *Stoch. Proc. Appl.* **16**, 55-69, (1983).

NEUROPHYSICS

Chaos, Boltzmann, Shannon and Electroencephalography

A. M. Albano[a] and R. B. Duckrow[b]

[a]*Department of Physics, Bryn Mawr College, Bryn Mawr, PA 19010-2899 USA*
[b]*Department of Neurology, Yale University School of Medicine, New Haven, CT 06520-8018, USA*

Abstract. Nonlinear dynamics has made possible the development of new tools for the quantitative analysis of electroencephalographic (EEG) signals. Some of the tools used in the initial applications required large quantities of noise-free, stationary data which are usually not available from biological systems. Information theoretic measures calculated using coarse-grained data are more appropriate for the analysis of these data. We review Nonlinear Dynamics and some nonlinear dynamical tools, introduce some Information Theoretic notions and use Mutual Information and Transfer Entropy to probe relationships among data streams in 19-channel scalp EEG recorded during three sleep stages – wakefulness, slow-wave sleep, and REM sleep.

Keywords: Nonlinear dynamics, chaos, information theory, sleep EEG
PACS: 87.19.1e, 87.19.lo, 89.70.-a, 05.45.-a, 05.45.Gg

INTRODUCTION

When Hans Berger first observed electrical activity in the human brain some 79 years ago [1], he dreamt that the study of these signals might eventually become ``a window on the mind." In the past three decades or so, with the parallel development of fast, affordable computers and of nonlinear dynamical techniques, the nature and amount of quantitative studies of the EEG have increased enormously. Yet, the fulfillment of Berger's dream seems as distant as ever.

The initial applications of nonlinear dynamics to the EEG made use of techniques that were meant for the analysis of large amounts of essentially noise-free, stationary data. These analyses also focused mainly on single EEG channels, in part because of the complexity of dealing with multi-channel data, and in part because of the expectation that information from one data channel carries with it information about all the other channels with which it is nonlinearly coupled [2]. There has since been a growing appreciation of the practical problems encountered when techniques meant for clean, stationary data are applied to inherently noisy and nonstationary signals such as electrical signals produced by biological systems. Similarly, there has been a realization that techniques which are appropriate for systems with relatively few degrees of freedom may not yield much useful information when applied to systems as complex and as multiply-connected as the brain (see, *e.g.*, [3]). These have led to the exploration, in recent years, of algorithms that are more robust against noise and nonstationarity, and of techniques for dealing with multichannel data.

CP1021, *Stochastic and Quantum Dynamics of Biomolecular Systems, Proceedings of the 5th Jagna International Workshop,* edited by C. C. Bernido and M. V. Carpio-Bernido

One way of dealing with imprecise data is *coarse graining*, a technique used by Boltzmann in the 1870's to formulate the kinetic theory of gases (see, *e.g.*,[4]). Instead of dealing with the positions and the momenta of the molecules, Boltzmann formulated his theory in terms of the probability, f_α, that the molecules have position and momenta within ranges specified by the index, α. Boltzmann's definition of entropy, $S = -k\Sigma_\alpha f_\alpha \log_2(f_\alpha)$, was eventually reflected in Shannon's definition of information. Information Theory, which Shannon developed in the 1940's [5], and which continues to evolve, provides some coarse-grained quantities for characterizing imprecise data that are proving to be useful in the study of biological as well as other "real-world" systems.

In this contribution, we include a brief review of some nonlinear dynamic and some information theoretic notions to introduce the use of two constructs that have gained some recent popularity: Mutual Information and Transfer Entropy [6] (also called Conditional Mutual Information [7].) We illustrate the use of these techniques by analyzing 19-channel human EEG data recorded during the first several minutes of sleep -- the first sleep cycle -- consisting of wakefulness, slow-wave sleep and rapid-eye-movement (REM) sleep.

DYNAMICS, LINEAR AND OTHERWISE

In its original context, dynamics meant Newton's $F=ma$. Its meaning has since metamorphosed so that a modern dynamical system is now more likely to be presented as a coupled set of first order differential equations,

$$dq_i / dt = f_i(q_1, q_2, \cdots, q_n), \quad i = 1, 2, \cdots, n, \tag{1}$$

which hides its Newtonian origins and barely disguises its relationship to Hamilton's equations of motion. The q_i's are a set of variables that specify the state of the system – they could be the positions and momenta of old, but they could as well be variables describing a laser, or the economy, or the atmosphere, or insect populations, or a brain. If the f_i's are linear in the q_j's, then we have linear dynamics, and the exact analytical solutions are well-known – they are exponentials. If, on the other hand, one or more of the f_i's are nonlinear in any of the q_j's, then we have nonlinear dynamics, and exact, closed-form solutions, in general, do not exist.

Example: Rössler

To illustrate some nonlinear dynamical concepts, we use a system of equations due to Rössler [8]:

$$\begin{aligned} dx/dt &= -(y+z) \\ dy/dt &= x + ay \\ dz/dt &= b + xz - cz \end{aligned} \tag{2}$$

Because of the nonlinearity, xz, in the third equation, there are no known closed form solutions. Numerical solutions for the parameter values, $a = 0.15$, $b = 0.20$, $c = 10$, are shown in Fig. 1.

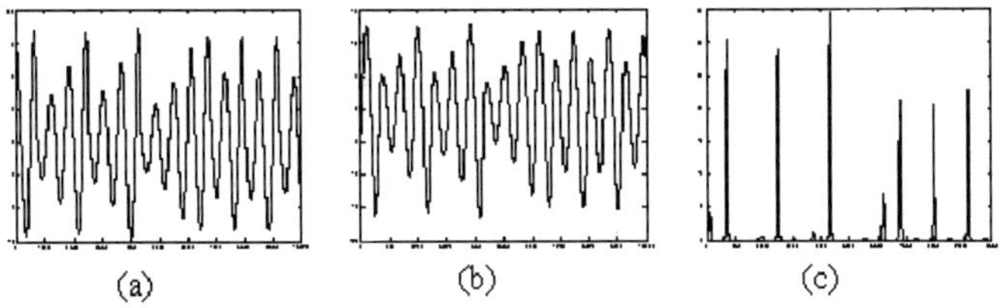

FIGURE 1. Numerical solutions of the Rössler equations (2) for the parameter values, $a = 0.15$, $b = 0.20$, $c = 10$. The x, y, and z variables are shown on the left, middle and right panels, respectively.

The system's time evolution can be better visualized by plotting its trajectory in the three-dimensional phase space spanned by x, y, and z (Fig. 2). After initial transients have decayed, the trajectory is confined to a bounded region in phase space called the system's *attractor*. The trajectory of Rössler's equation does not repeat itself. In other words, it is *deterministic, bounded, aperiodic motion*, that is, it is *chaotic* [9]. While the behavior of the individual variables seem to be random, the attractor itself shows some order.

The Experimental Situation: Embedding

The foregoing raises the possibility that seemingly random data measured from complex systems might be used to construct "orderly" phase space trajectories. This means that, given a dynamical system characterized by variables, q_1, q_2, …, q_N, we would need to measure sequences of values, *time series* so-called, of all the variables; use these variables to plot phase space trajectories; and then use geometric, or dynamic, or topological properties of the trajectories to characterize the system. That is, if we know what all the variables are, to begin with, and if we are in a position to measure them. When dealing with systems as complicated as the atmosphere, or the economy, or the human brain, this is hardly possible.

Fortunately, it is, in principle, not necessary to know or to measure all of the system's variables. A remarkable theorem, the *Embedding Theorem*, proved at about the same time independently by Takens [10] and by Mañé [11], says that the system's phase space trajectory can be reconstructed using a time series of only one of the system's variables. We demonstrate it by using the Rössler system once again.

FIGURE 2. The trajectory of the Rössler system in the phase space spanned by (x,y,z).

The "natural" phase space of the Rössler equations (Eq. 2) is that spanned by the variables, x, y. and z, as shown in Fig.2. It is, however, possible to eliminate y and z and to rewrite the three equations as a single third order differential equation for x,

$$\dddot{x} = \ddot{x}(c-x) + \dot{x}(ax)c - x - a) - 1) - x(c - x - a)$$
$$-\left(\frac{\ddot{x} + a\dot{x} - x - b}{c - x}\right)\left(\dot{x} + x + c - a^2\right). \tag{3}$$

The "natural" phase space of Eq. (3) is that spanned by x, $\dot{x} = dx/dt$, and $\ddot{x} = d^2x/dt^2$. The trajectory in this space is shown in Fig. 3(a). Compared to Fig. 2, the figure is rotated, and perhaps distorted, but the similarity is obvious.

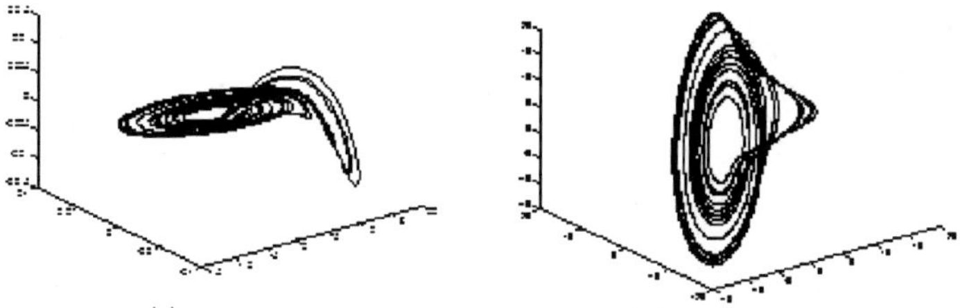

(a) (b)

FIGURE 3. The trajectory of the Rössler system (a) in the phase space spanned by x, dx/dt, and d^2x/dt^2, (b) in the phase space spanned by coordinates $(x(k\tau), x((k\text{-}50)\ \tau), x((k\text{-}100)\ \tau))$.

In a situation where all one has is a time series for x, *i.e.*, $x(\tau)$, $x(2\tau)$, …, $x(k\tau)$, … that is, a sequence of measurements of x separated by the sampling time, τ, the derivatives at time, $k\tau$, can be approximated by

$$\begin{aligned} dx/dt\big|_k &\approx \frac{x(k\tau) - x\big((k-1)\tau\big)}{\tau}, \\ d^2x/dt^2\big|_k &\approx \frac{x\big((k+1)\tau\big) - 2x(k\tau) + x\big((k-1)\tau\big)}{\tau^2}, \end{aligned} \tag{4}$$

etc. The values, $x(k\tau)$, $x((k+1)\tau)$, ..., $x((k+n)\tau)$ contain approximately the same information as x, dx/dt, ... , d^nx/dt^n. Instead of using a space spanned by x and its derivatives, one may use a space spanned by the "time-delay coordinates" ($x(k\tau)$, $x((k-1)\tau)$, $x((k-2)\tau)$, ..., $x((k-n)\tau)$. Using time delay coordinates instead of approximations of the derivatives has the following added advantage: if the $x(k\tau)$ are measured values of a continuous variable, and if τ is sufficiently small, then successive values of $x(k\tau)$ are likely to be close to each other so that their differences, which are used to calculate derivatives, could be of the same order of magnitude as the measurement errors themselves. Using time delay coordinates instead of derivatives involves dealing with more precise information.

An n-dimensional vector with time-delay coordinates is of the form

$$X(k) = \left(x(k\tau),\ x((k-1)\tau),\ x((k-2)\tau)\ ,\ \cdots,\ x((k-n-1)\tau)\right). \tag{5}$$

Takens and Mañé showed that for a dynamical system with a d-dimensional attractor, using $n > 2d$ delay coordinates reconstructs the trajectory in a manner that preserves properties of the attractor that are invariant under smooth coordinate changes. Sauer *et al.* [12] later showed that $n = 2d$ may be enough. This reconstruction is called an *embedding*; the vectors used in the embedding (Eq. 5) are the *embedding vectors*.

Fig. 3(b) shows a phase space trajectory of the Rössler system using the coordinates ($x(k\tau)$, $x((k\text{-}50)\ \tau)$, $x((k\text{-}100)\ \tau)$). As in the case of Fig. 3(a), we see that this is a rotated version of the attractor shown in Fig. 2.

The use of delay coordinates that are separated from each other by 50τ -- *i.e.*, using an *embedding lag* of 50 rather than just 1, or 17 say, illustrates a practical problem with embedding. If the lag is too small, then successive values of x are likely to be close to each other, and the embedded points would all be close to the main diagonal of the phase space. This makes it difficult, if not impossible, to sort out details of the reconstructed trajectory. The problem gets even worse if the data are imprecise. On the other hand, if the lag is unduly large, then the components of the embedding vectors may be uncorrelated, resulting in a trajectory that jumps randomly in the embedding space.

Another practical embedding problem concerns the choice of the dimension of the embedding space. The Embedding Theorem gives a lower limit for this number if the dimension of the attractor being reconstructed is known, but determining this dimension could well be one object of the analysis. Developing procedures to find optimal values of these two quantities, the embedding lag and the embedding dimension, continues to be an active area of research (see, *e.g.*, [13]).

Dimensions, Fractals, and Strange Attractors

An attractor reconstructed using embedding vectors of the form of Eq. (5) is a swarm of points in the n-dimensional embedding space. The attractors of *linear* dynamics are simple. A damped oscillator eventually comes to rest – its attractor is a point, an object of dimension 0. The attractor of an undamped oscillator in the phase space spanned by its position and momentum is an ellipse – an object of dimension 1; that of two coupled oscillators with incommensurate frequencies is a 2-torus, an object of dimension 2. Almost all chaotic attractors, however, have non-integer dimensions – they are *fractals*. This is why Ruelle and Takens [14] referred to them as *strange attractors*.

Probably the most widely-used quantity for estimating the dimension of a set of points uses the *Correlation Integral*, $C_m(r)$, which may be defined as the fraction of distances between the points that do not exceed r. Grassberger and Procaccia [15] have shown that for a sufficiently large embedding dimension, m, and small enough distances, there is a range of r*'s* in which the correlation integral scales as r^{D_2} where D_2 is the *correlation dimension* of the set of embedded points.

The relative ease with which D_2 can be calculated led to a number of claims, in the mid-1980's, that the brain displays low-dimensional chaos as indicated by small, noninteger values of D_2 and similar dimension estimates (see, *e.g.*, [2]). It soon became evident, however, that fractal dimension is not a sufficient condition for chaos. Data sets of filtered random noise, for instance, can appear to have small fractal dimensions [16]. A study by Theiler and Rapp [17] suggested that many earlier claims of low-dimensional chaotic behavior in the brain based on fractal dimension calculations may not withstand more careful scrutiny.

Lyapunov Exponents

Lyapunov exponents are a measure of the exponential rate of divergence, or convergence, of initially nearby trajectories in phase space. We may introduce them informally as follows: Consider a set of trajectories of a dynamical system in m dimensions. Let the initial points of these trajectories lie in a small hypersphere of radius, r_0. After some time, δt, the trajectories will have evolved so that their end points are no longer all inside a sphere of radius, r_0. Some of the trajectories may have moved closer to each other, others farther apart so that their end points will now be contained in an ellipsoid. The lengths of the semi-axes of the resulting ellipsoids are averaged over all possible centers of the sphere of initial points, and these are expressed as $r_i = r_0\exp(\lambda\delta t)$. The quantities, λ_i, are the Lyapunov exponents of the motion. If at least one of the λ_i's is positive, the motion is chaotic.

Calculating Lyapunov exponents accurately can be more problematic [18] and may require even larger amounts of noise-free, stationary data than calculating dimensions [19]. Large amounts of noise-free stationary data, however, are not the kinds of data obtainable from biology or from other real world systems. For these systems, there is a clear need for techniques which are more robust against imprecise data and noise, and which are more capable of dealing with relatively short data streams to try to avoid the complications of nonstationarity.

COARSE GRAINING, MISSING INFORMATION, AND INFORMATION THEORETIC MEASURES

Boltzmann and Shannon

Imprecise or noisy information was not one of the problems that confronted Boltzmann when he formulated kinetic theory. What he confronted was the problem of having too much precise information. His use of coarse-graining to solve his problem manages to deal also with ours.

The phase space of a gas of N point particles is $6N$-dimensional, spanned by $3N$ position and $3N$ momentum coordinates. A point in this space represents precise values of the positions and the momenta of all the particles. Boltzmann partitioned this space into cells with dimensions that are small compared to the dimensions of the system, but large enough to contain many representative points. Rather than dealing with the precise location of representative points, he dealt only with the phase space cells in which the points are located. The theory is formulated in terms of f_α, the probability that a representative point is found in the cell labeled by the index, α -- that is, a probability distribution function (PDF) of a discrete random variable. Boltzmann's statistical definition of entropy, $H_B = -k\sum_\alpha f_\alpha \log_2 f_\alpha$, justifiably well-known for its role in kinetic theory, is also a very convenient way of characterizing, by a single quantity, the distribution of points in the coarse-grained phase space.

Shannon was concerned with information about messages encoded by the use of finite alphabets. His definition of "missing information," now usually referred to as *Shannon entropy,*

$$H = -\sum_\alpha f_\alpha \log_2 f_\alpha, \tag{6}$$

is the essentially same as Boltzmann's, except that now, f_α is the relative frequency with which the alphabetic symbol represented by α appears in the message. H is, of course dimensionless, but when the logarithm is calculated to base-2, it is customarily given the artificial unit of 'bits.'

Some Information Theoretic Functionals

We define some functionals that have become useful in characterizing relationships between data sequences from complex systems. Let A and B be two discrete random variables with PDF's f_α and f_β, respectively. The *joint PDF*, the probability of the simultaneous occurrence of α and β is $f(\alpha,\beta)$, and the *conditional PDF*, the probability that α occurs given that β occurs, is $f(\alpha|\beta)$. The last two are related by

$$f(\alpha,\beta) = f(\alpha\,|\,\beta) f(\beta). \tag{7}$$

The *joint entropy*, $H(A,B)$, of A and B is

$$H(A,B) = -\sum_\alpha \sum_\beta f(\alpha,\beta) \log_2 f(\alpha,\beta). \tag{8}$$

If A and B are independent, then $f(\alpha,\beta) = f(\alpha)f(\beta)$ and $H(A,B) = H(A) + H(B)$

The *conditional entropy*, $H(A,B)$, of A given B is

$$H(A|B) = -\sum_{\alpha}\sum_{\beta} f(\alpha,\beta)\log_2 f(\alpha|\beta) \tag{9}$$

The *Mutual Information*, $M(A,B)$, the amount of information common to A and B is defined as

$$M(A,B) = H(A) + H(B) - H(A,B). \tag{10}$$

$M(A,B)$ is a measure of the departure of the joint entropy, $H(A,B)$, from the assumption that A and B are independent, for which case $M(A,B) = 0$. It is a generalization of the usual measures of cross-correlations such as Pearson's r (see below.) $M(A,B)$ is capable of detecting nonlinear correlations while r is not.

In terms of the PDF's, *Transfer Entropy*, $T_{B\to A}$, the amount of information transferred from B to A is [6]

$$T_{B\to A} = \sum_{k} f(\alpha_{k+1},\alpha_k,\beta_k)\log_2 \frac{f(\alpha_{k+1}|\alpha_k,\beta_k)}{f(\alpha_{k+1}|\alpha_k)}. \tag{11}$$

It is a measure of the influence of the sequence, B, on the evolution of A. Palus [7] showed that this is equivalent to the *conditional mutual information,* the mutual information of $A_1=\{\alpha_1, \alpha_2, \ldots,\alpha_{N-1}\}$ and $A_2=\{\alpha_2, \alpha_3, \ldots,\alpha_N\}$ given $B = \{\beta_1, \beta_2, \ldots, \beta_{N-1}\}$, defined by $M(A_1;A_2|B) = H(A_1|B) + H(A_2|B) - H(A_1;A_2|B)$. In the following, we use Schreiber's formulation.

INFORMATION THEORETIC MEASURES AND SLEEP EEG

In this Section, we use information theoretic measures to study relationships among 19-channel scalp EEG measured from six subjects recorded during wakefulness, slow-wave sleep and REM sleep. Wakefulness is actually when the subject is about to fall asleep, slow-wave sleep is associated with deep sleep and REM sleep is associated with dreaming when the subject shows much rapid-eye movements, hence the name. Some of the subjects did not provide data for all sleep stages. The data include five episodes of wakefulness, six of slow-wave sleep and four of REM sleep. The data segments used in this study have been chosen to be free of artifacts.

Fig. 4 shows segments of one set of 19-channel EEG for each of the three sleep stages. The symbols on the cartoon of the head identify the locations of the scalp electrodes used in the recording. The symbols consisting of a letter and a number, *e.g.,* F3, are standard notations used by encephalographers to identify well-defined locations on the scalp. The numbers, 1, …, 19 are just introduced for convenience.

Fig. 5 shows some one-channel details of the wave forms for the three sleep stages. The major qualitative features of these are well known to encephalographers who use these along with other criteria to classify sleep stages. Fig. 6 shows PDF estimates using 1000-point epochs. Corresponding to the increase in the sharpness of the peaks of the PDF's as one goes from awake to slow-wave to REM sleep, the Shannon Entropy decreases from 9.00 bits for awake, 8.60 bits for slow-wave sleep, and 8.42 bits for REM sleep.

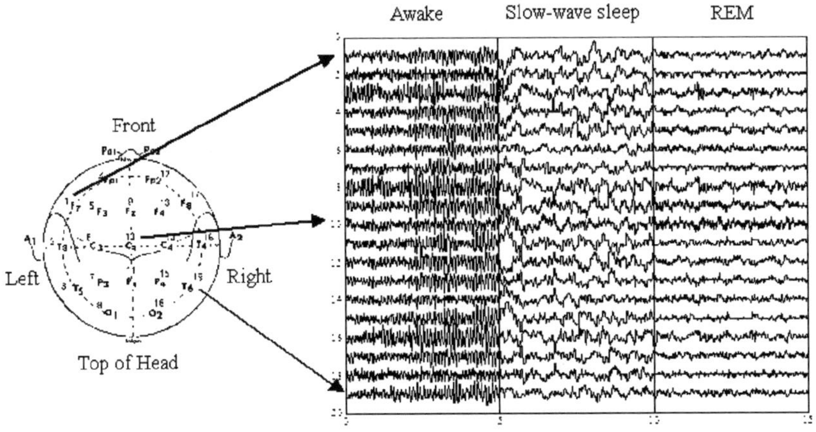

FIGURE 4. Electrode locations on the scalp and segments of EEG recorded during wakefulness, slow-wave sleep, and REM sleep.

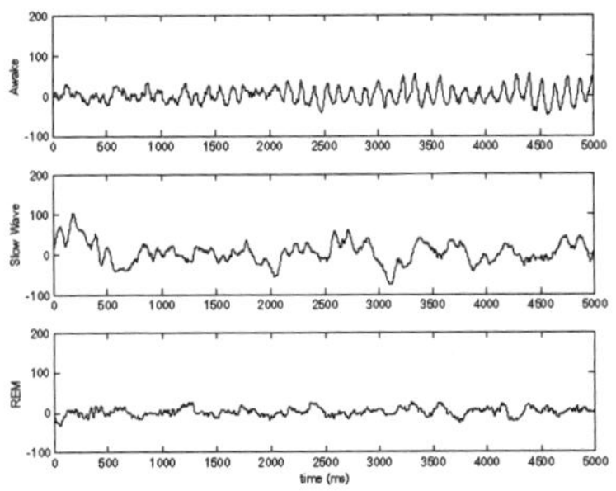

FIGURE 5. Details of single channel recordings during (top) wakefulness, (middle) slow-wave sleep, and (bottom) REM sleep. The data were recorded using an average reference at 200 samples/second. The recording bandpass was 0.5 – 100 Hz.

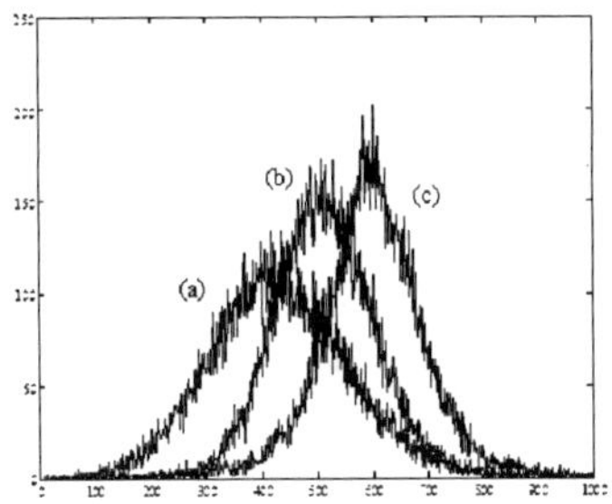

FIGURE 6. PDF estimates using 1000-pt epochs. The corresponding Shannon Entropies are (a) Awake, $H = 9.00$ bits, (b) REM, $H = 8.42$ bits, (c) Slow- wave, $H = 8.60$ bits

Linear and nonlinear correlations

Entropy calculations can provide quantitative confirmation of some of the structural features of single-channel wave forms that may already be discernible by visual inspection. The greater utility of information theoretic measures lies in their ability to detect relationships between pairs of data channels.

A widely used measure of the linear correlation of two data sequences, $X = \{x_1, x_2, \ldots, x_N\}$ and $Y = \{y_1, y_2, \ldots, y_N\}$, is Pearson's r:

$$r = \frac{1}{N}\sum_{i=1}^{N}\frac{(x_i - \bar{x})(y_i - \bar{y})}{\sigma_x \sigma_y}. \tag{12}$$

To compare the linear cross-correlations of the data, we averaged r for ten 1000-point segments randomly chosen from all 171 distinct channel pairs of each sleep stage. This was done for all sleep stage data available from all subjects. These averages are shown in Fig. 7 which shows that the linear correlations for the three sleep stages do not have statistically significant differences.

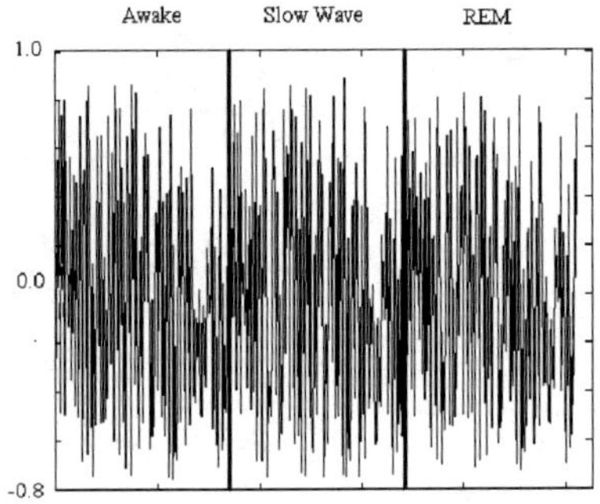

FIGURE 7. Linear correlations for 171 channel pairs per sleep stage. Each point is the value of Pearson's r averaged over ten randomly chosen 1000-point segments and over all subjects.

In contrast to linear correlations, nonlinear correlations, as measured by mutual information shows distinct differences between the sleep stages. Fig. 8 shows the mutual information averaged over all channel pairs and all subjects calculated using the same data epochs used to obtain Fig.7. On average, slow wave sleep showed the highest level of mutual information, followed by wakefulness, and then by REM sleep.

Transfer Entropy and Information Transfer

Mutual Information is a measure of the information that two data sequences have in common. If the two sequences being compared are driven by a third but are otherwise non-interacting, or if one channel drives the other, mutual information is not capable of distinguishing the difference, but transfer entropy is. Fig. 10 contrasts these two measures using data from *one subject* (see the figure caption for more detail). The quantity represented in the lower panels is the *Net Transfer Entropy,* $T_{j\to i} - T_{i\to j}$.

During wakefulness (top left), for instance, the mutual information of channel 1(F7) and the other channels is relatively small as indicated by the blue horizontal band at row 1, but there is considerable Net Transfer Entropy from all the other channels to channel 1 as indicated by the orange and red band at this row. The same is true of channels 2 (T3), 17(F8) and 18(T4). In the case of the latter two, Net Transfer entropy to these channels is indicated by the blue bands at columns 17 and 18.

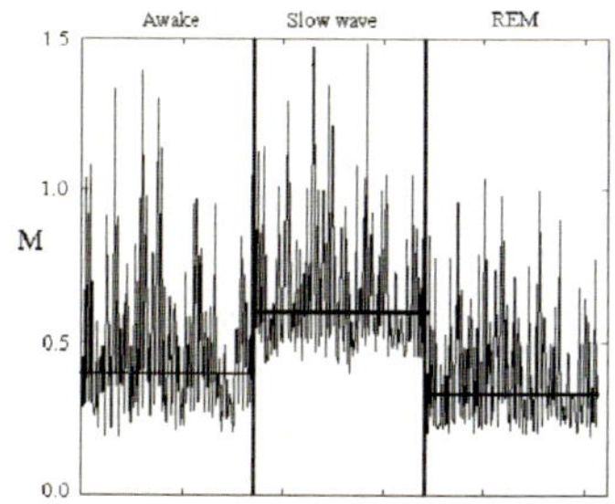

FIGURE 8. Pairwise mutual information using the same data epochs used to calculate r in Fig. 7. The horizontal lines for each sleep stage show the average values.

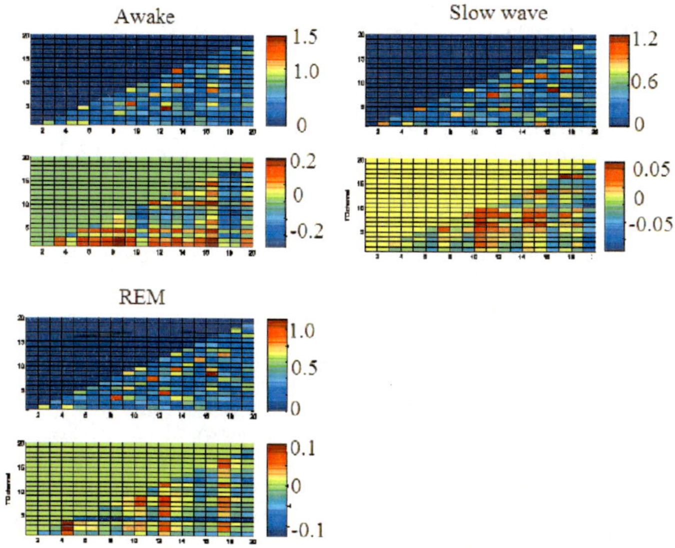

FIGURE 9. Mutual Information (upper panels) and Net Transfer Entropy (Lower panels) for the three sleep stages. Rows are numbered from bottom to top, columns from left to right. In the upper panels, the color of cell at row=i, column=$j > i$ is the Mutual Information of channels i and j; in the lower panels, it is the Net Transfer Entropy from channel j to channel i. Note the differences in scale as indicated by the color bars.

It remains to be seen if the patterns of Mutual Information and of Transfer Entropy that we have shown here are true in general, or if they are meaningful, or if they are consistent with known physiology. It would even be more interesting if they provide previously unknown information about processes in the brain during the various stages of sleep. In any event, even if they do not yet provide a way of opening the "window on the mind" that Berger sought, they may at least provide yet one more set of tools to help with the quest.

ACKNOWLEDGMENTS

We gratefully acknowledge the help of Paul Rapp and Chris Cellucci. AMA's participation in the 5th Jagna International Workshop was sponsored by the Balik Scientist Program of the Department of Science and Technology, Republic of the Philippines.

REFERENCES

1. H. Berger,*Arch. Psychiat. Nervenkr*. **87**, 527 (1929).
2. P.E. Rapp, I.D. Zimmerman, A.M. Albano, G. C deGuzman, N. N Greenbaun, and T.R. Bashore, "Experimental studies of chaotic neural behavior: cellular activity and electroencephalographic signals," in *Nonlinear Oscillations in Biology and Chemistry*, Lecture Notes in Biomathematics, edited by H. G. Othmer, Berlin: Springer-Verlag (1985), pp. 175-205; G. Mayer-Kress, and S. P. Layne, *Ann. N. Y. Acad. Sci.*, **504**, 62 (1987).
3. A. M Albano and P. E. Rapp, "On the reliability of dynamical measures of EEG signals," in *Nonlinear Dynamical Analysis of the EEG*, edited by B. H. Jansen and M. E. Brandt, Singapore: World Scientific (1993), pp. 117-139
4. P. and T Ehrenfest. *The Conceptual Foundations of the Statistical Approach in Mechanics*, translated by M. J. Moravcsik, Ithaca, NY: Cornell University Press, 1959.
5. O. Shannon, and W. Weaver, *The Mathematical Theory of Communication*, Urbana, IL: University of Illinois Press, 1949.
6. T. Schreiber, *Phys. Rev. Lett.* **85**, 461 (2000).
7. T. M. Cover and J. A. Thomas, *Elements of Information Theory* , New York :Wiley-Interscience, 1991; K. Schindler-Hlav´ackov´a, M. Palus, M. Vejmelka, and J. Bhattacharya, *Phys. Rep*. **441**, 1 (2007).
8. O. E. Rössler, *Phys. Lett.* **57A** 397 (1979).
9. T.-Y. Li, and J. A. Yorke, *Amer. Math. Monthly*, **82**, 985 (1975),.
10. F. Takens, in *Dynamical Systems and Turbulence,* Lecture Notes in Mathematics, **898**, edited by D. A. Rand and L. S. Young, New York: Springer-Verlag (1980).
11. R. Mañé, in *Dynamical Systems and Turbulence,* Lecture Notes in Mathematics, **898**, edited by D. A. Rand and L. S. Young, New York: Springer-Verlag (1980).
12. T. Sauer, J. A. Yorke, and M. Casdagli, *J. Stat. Phys*., **65**, 579 (1991).
13. H. D. I. Abarbanel, *Analysis of Observed Chaotic Data*, New York: Springer-Verlag (1996); T. H. Buzug and G. Pfister, *Physica D*, **58**, 127 (1992); C. J. Cellucci, A. M. Albano, and P. E. Rapp, *Phys. Rev. E* **,67**, 066210 (2003).
14. D. Ruelle, and F. Takens, *Commun. Math. Phys.* **20**, 167 (1971).
15. P. Grassberger, and I. Procaccia, *Phys. Rev. Lett.,* **50**, 346 (1983),.
16. A.R. Osborne, and A. Provenzale, *Physica D*, **35**, 357 (1989); P. E. Rapp, A. M. Albano, T. I. Schmah, and L. A. Farwell, *Phys. Rev. E*, **47**, 2289 (1993).
17. J. Theiler and P. E. Rapp, *Electroencephalogr. Clin. Neurophysiol.,* **98**, 213 (1996).
18. K. J. Blinowska and M. Malinowski, *Biol. Cybern.,* **66**, 159 (1991).
19. J. P. Eckmann and D. Ruelle, *Physica D*, **56**, 185 (1992).

Beneficial role of noise in artificial neural networks

Christopher Monterola*,†, Caesar Saloma* and Martin Zapotocky†

*National Institute of Physics, University of the Philippines
1101 Diliman Quezon City
†Max-Planck Institut für Physik Komplexer Systeme
Nöthnitzerstrasse 38, 01187, Dresden, Germany

Abstract. We demonstrate enhancement of neural networks efficacy to recognize frequency encoded signals and/or to categorize spatial patterns of neural activity as a result of noise addition. For temporal information recovery, noise directly added to the receiving neurons allow instantaneous improvement of signal-to-noise ratio [Monterola and Saloma, Phys. Rev. Lett. 2002]. For spatial patterns however, recurrence is necessary to extend and homogenize the operating range of a feed-forward neural network [Monterola and Zapotocky, Phys. Rev. E 2005]. Finally, using the size of the basin of attraction of the networks learned patterns (dynamical fixed points), a procedure for estimating the optimal noise is demonstrated.

Keywords: Noise, Stochastic Resonance, Neural Networks
PACS: 84.35.+i, 72.70.+m, 89.75.Fb, 02.50.Ey

INTRODUCTION

The role of noise in neural processing has received significant attention in recent years, particularly in the context of stochastic resonance (SR) [3]. In the SR mechanism, noise is used in a counter-intuitive way to permit detection of subthreshold signals of a particular frequency by a thresholding neuron. Noise-enhanced sensitivity has been observed previously in a variety of biological neural systems like: 1) Periodically-stimulated crayfish neurons [4], 2) Information encoding in primary auditory nerve of squirel monkey [5], 3) Prey targetting of paddle fish [6] and cricket [7], 4) Human visual perception [8], and 5) Heart rate compensation for blood pressure regulation [9].

Here, we show that the sensitivity of a network of integrate-and-fire neurons that already learned to recognize specific features of strong (above-threshold) disturbances in a noise-free environment, is enhanced by ambient noise without re-training. We discussed here the two most important protocols for pattern recognition: 1) recovering *frequency encoded information* where signal's signature lies in the temporal structure of the stimulus and 2) recovering *spatial encoded information* in which the stimulus is mapped into the pattern of activity across a set of neurons at a specific time.

NOISE-DRIVEN RECOVERY OF TEMPORAL SIGNAL

The response Ψ_N^q of our single-output network is described by: $\Psi_N^q = f_o(\sum_h d_{ho}^q y_h^q)$ where:$y_h^q = f_H(\sum_m w_{mh}^q x_m^q)$, $w_{mh}^q(d_{ho}^q)$ represents the interconnection weight from the

CP1021, *Stochastic and Quantum Dynamics of Biomolecular Systems*,
Proceedings of the 5th Jagna International Workshop, edited by C. C. Bernido and M. V. Carpio-Bernido

mth-input to the hth-hidden neuron (node) immediately after the qth iteration, $f_H(x)$ and $f_o(x)$ are the hidden and output activation functions respectively, $m = 1,2,...,M$; $h = 1,2,...,H$. The network samples a sinusoid disturbance $i(x)$ via its M input nodes, into a sequence $\{i(x_m)\}$ that satisfies the Nyquist sampling criterion where: $x_m = x_{m-1} + \Delta x$, and x is an independent variable. The m-th input node outputs a "1" if $i(x_m) > A_{th} = 0.5$. Otherwise, it outputs a zero. The response of the hidden and output nodes is described by: $f_H(n) = f_o(n) = [1 + exp(-n)]^{-1}$.

Neural interconnection strengths are established by minimizing a non-negative cost function $E^q = 0.5(\Psi_N^q - \Psi_t^q)^2$, where Ψ_t is the desired response which is equal to 1.0 if: 1) $x(t_m) > A_{th}$, and 2) Frequency f of $i(x)$ is equal to a pre-determined value f_r. A mature network outputs a "1" if the f-value of an arbitrary disturbance is f_r. Otherwise, it outputs a zero. Learning proceeds using the standard gradient-descent backpropagation method which is the most standard and succesful learning algorithm for neural networks (see appendix A for details).

The disturbance is described by: $i(x) = (A/2)[sin(2\pi x) + 1]$, where: $0 \leq A$ (7-bit precision) ≤ 1. To satisfy the Nyquist criterion, only $(M/2)$ f-values are possible for $i(x)$, i.e. $f = 1/M, 2/M, ..., 1/2$; where $2^6 M$ is the total number of possible network input states. For our experiments, we utilize a network with $M = 50$ and $H = 10$ ($2^6 M = 3200$, $\eta = 0.001$) - others within: $6 \leq H \leq 35$, were also highly trainable. Our training set consists 2560 elements taken from all the 128 possible $x(t)$ values at $f = f_r$, plus 2432 other randomly-chosen $x(t)$ values for $f \neq f_r$. The network learns perfectly (100% accuracy for both training and test sets after $q = 500$) at any f_r except at $f_r = 1/2$ which is the cut-off frequency of sampling and hence prone to aliasing [10].

Performance of trained network as a frequency detector

Figure 1a plots the output Ψ_N^{500} as a function of A ($f_r = 10, A = 0.4$) for the test set. Without noise, the network recognizes only strong $i(x)$'s with $A > 0.5$ and $f = 10$. Sub-threshold $i(x)$'s are unrecognized even at $f = 10$.

With additive ambient noise, weak $i(x)$'s with $f = 10$ become recognizable by the trained network. Fig 1b plots the success rate S as a function of noise variance σ^2 for uniform-white (filled squares) and Gaussian (squares) noise. The $S(\sigma^2)$ plot is nonlinear for the dithered disturbance: $i(x) + n_\sigma(x)$. S increases with σ^2 with the detection onset occurring at $\sigma^2 \approx 0.1$(white noise), which is expected since $A = 0.4$ and $A_{th} = 0.5$. For $\sigma^2 > 0.44$, S decreases with σ^2 such that $S \approx 0$ at $\sigma^2 > 1.0$.

We compare the network results with those from a simple threshold system using Gammaitoni's analysis [11]. For a given noise type, the peak location of σ^2 for the network ($\sigma \approx 0.55$) is very near that of a simple system ($\sigma \approx 0.52$). White-noise dithering yields larger S values than Gaussian noise, for both the network and simple system. However, the network yields sharper $S(\sigma^2)$ curves with onsets that occur at larger σ^2, indicating that it does not operate exactly like a simple system. Without loss of generality, we present results only from white-noise dithering because it yields larger S values and its $S(\sigma^2)$ curve falls off faster with increasing σ^2. Similar behavior of the $S(\sigma^2)$ curve is also found with other f_r values except at $f_r = 1/2$. On the other hand,

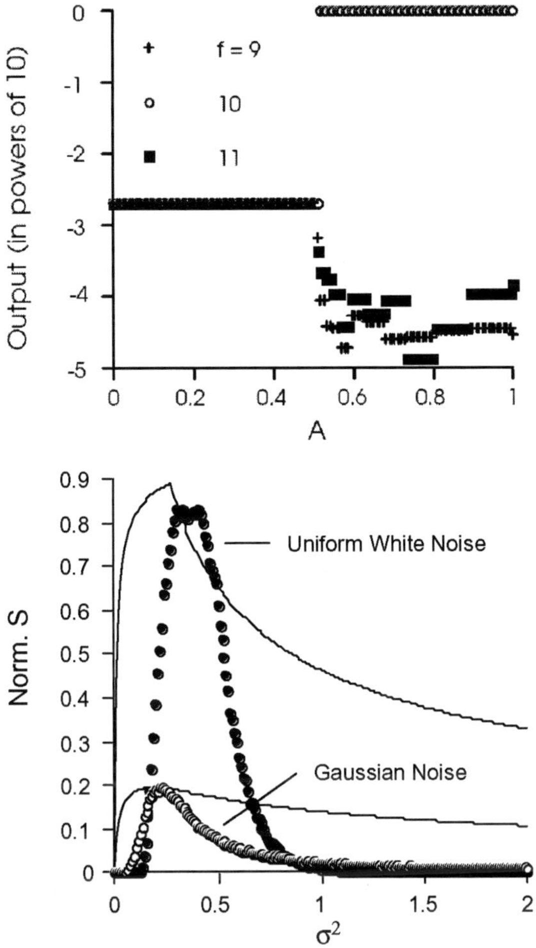

FIGURE 1. Performance of mature network ($f_r = 1/5$). a)Ψ_N^{500} vs A ($\sigma^2 = 0$) and $f = 9/50$, $10/50$, $11/50$; and b) Normalized $S(\sigma^2)$ for white noise (filled squares) and Gaussian noise (squares) for $A = 0.4$ and 5000 trials. With or without noise, trained network recognizes only $i(x)$'s with $f = 1/5$. Also in Fig 1b are the predicted $S(\sigma^2)$'s (solid curves) for a simple threshold system.($A_{th} = 0.5$)

noise effects were unobserved in non-f_r values indicating that learning was unaffected by ambient noise.

Performance of trained network for different signal strength

Noise-aided improvement of the detection limit is evaluated at different A values ($0.10 \leq A \leq 0.45$). Figure 2 shows that detection begins at $\sigma^2 \approx (0.5 - A)$. Within:

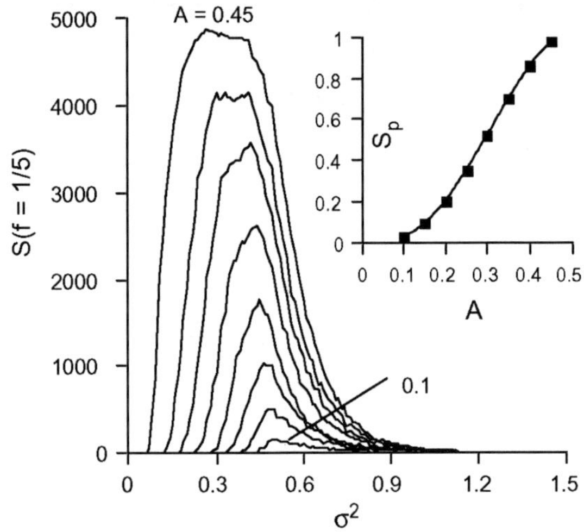

FIGURE 2. Performance of trained network ($f_r = 1/5$, 5000 trials). $S(\sigma^2)$ for $f = 1/5$ and $A = 0.45$ (topmost curve), $0.4, 0.35, 0.3, 0.25.0.2, 0.15$, and 0.1 (lowermost curve). Inset: S_p vs A

$0.10 \leq A \leq 0.45$, the peak value S_p of σ^2 decreases with A according to (best-fit curve): $S_p = -13.727A^4 - 9.817A^3 + 17A^2 - 2.52A + 0.127$, with an average standard deviation of ± 0.0049. Via the Newton-Raphson method, the smallest detectable A by the trained network is determined at: 0.08 ± 0.0008, representing a sensitivity improvement of 625%.

Trained network as a lowpass, bandpass and highpass frequency filter

Feature-extraction involves the recognition of a specific set of f-components in a multifrequency disturbance [12]. The effect of ambient noise is tested on a trained network that could perform low-, band-, or high-pass filtering of strong disturbances. Low-pass filtering permits the determination of the relative strengths of i(x) since its energy is mostly in the low-frequency components. Fine (low-energy) details of a disturbance are contained in its high-frequency band.

A single-output network ($M = 50, H = 10$) is trained to identify a set of equi-amplitude $i(x)$'s with the following f-values ($A = 0.4$): Set I ($fr = 1/50, 10/50, 21/50, 23/50$), Set II ($1/50 \leq fr \leq 1/10$), Set III ($1/5 \leq fr \leq 14/50$), and Set IV ($21/50 \leq fr < 25/50$). In any set, a network could be trained rapidly to yield perfect results in less than 500 iterations.

In Fig 3a are the corresponding $S(\sigma^2)$ for Set I, which all have onsets at: $\sigma^2 \approx 0.1$. Their peaks are within: $0.3 \leq \sigma^2 \leq 0.4$, where: $< S_p >= 4211 \pm 290$. Ambient noise is beneficial only for Set I frequencies - weak $i(x)$'s with different f-values remain hard to detect (see inset S plots). The S_p's in the inset plots are typical of all non-Set I frequencies. The standard deviation of $< S_p >$ for the Set I frequencies is well

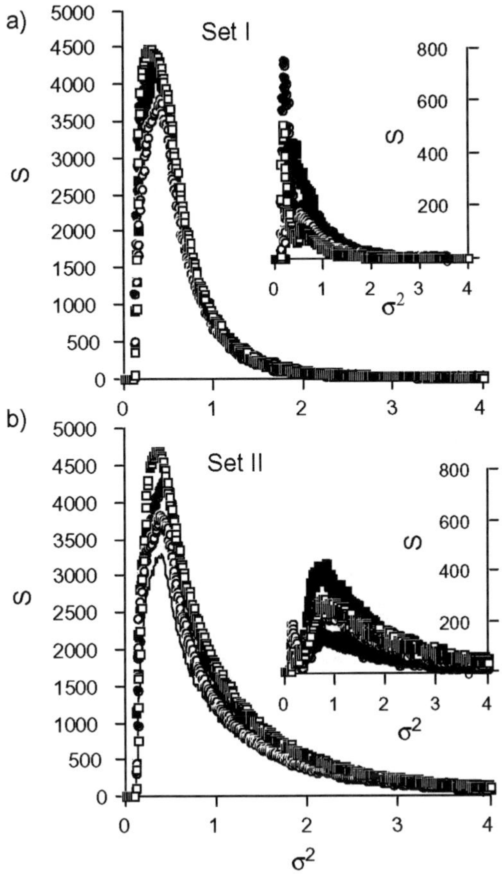

FIGURE 3. $S(\sigma^2)$ for $A = 0.4$ (5000 trials). a) Selected f-values, $f_r = 1/50$ (squares), $1/5$ (circles), $21/50$ (filled squares), and $23/50$ (filled circles); b) Low-frequency band, $f_r = 1/50$ (squares), $2/50$ (circles), $3/50$ (filled squares), $4/50$ (filled circles) and $1/10$ (solid line). Inset of Fig 3*a* are plots for $f = 3/50$ (squares), $11/50$ (circles), $20/50$ (filled squares), and $24/50$ (filled circles). Inset of Fig 3*b* are plots for $f = 11/50$ (squares), $12/50$ (circles), $13/50$ (filled squares), $14/50$ (filled circles).

within the $< S_p >$ values yielded by non-Set I signals. In Fig 3*b* are the associated $S(\sigma^2)$'s for Set II, with $< S_p >= 3983.2 \pm 538$. Again, noise is beneficial only within the predetermined low-frequency band (see inset plots). In Fig 4*a* are the associated $S(\sigma^2)$'s for Set III, where $< S_p >= 4201 \pm 209$. Noise is favorable only for Set III frequencies (see inset S plots for other values). In Fig 4*b* are corresponding $S(\sigma^2)$'s for Set IV where: $< S_p >= 4303 \pm 397$.

Figures 3 and 4 illustrate that for the four sets of f_r's, the resulting $S(\sigma^2)$'s have the same detection onset at $\sigma^2 \approx 0.1$, and range $(0.3 \leq \sigma^2 \leq 0.4)$ for their peaks. The benefits of noise are therefore realized uniformly in various filtering schemes.

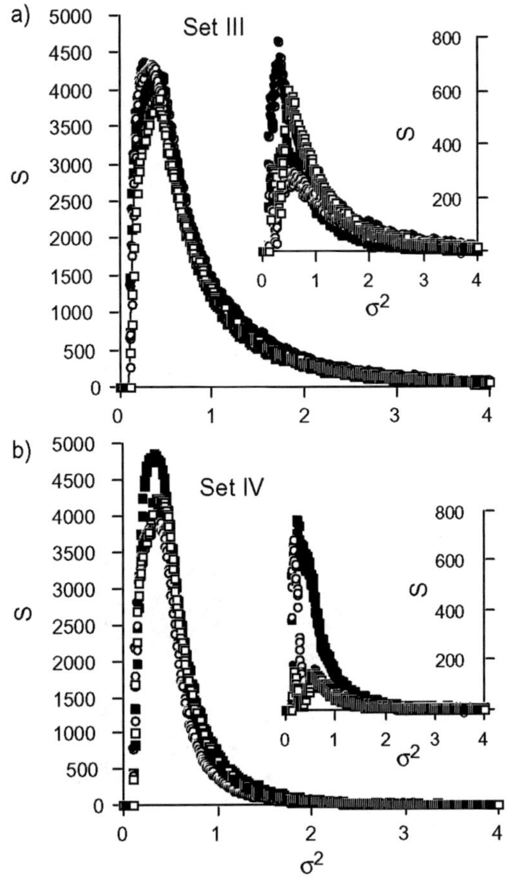

FIGURE 4. $S(\sigma^2)$ for $A = 0.4$ (5000 trials). a) Mid-frequency band, $f_r = 10/50$ (squares), $11/5$ (circles), $12/50$ (filled squares), $13/50$ (filled circles) and $14/50$ (solid line); b) High-frequency band, $f_r = 21/50$ (squares), $22/50$ (circles), $23/50$ (filled squares), and $24/50$ (filled circles). Inset of Fig 4*a* are plots for $f = 8/50$ (squares), $9/50$ (circles), $15/50$ (filled squares), and $16/50$ (filled circles). Inset of Fig 3*b* are plots for $f = 11/50$ (squares), $12/50$ (circles), $13/50$ (filled squares), $14/50$ (filled circles).

Network response variations among the pre-selected f_r's are within network's residual performance on other non-f_r values.

Above-threshold disturbances

Figures $1-4$ reveal that ambient noise of the right strength enables the detection of weak $i(x)$'s by a mature network without re-training. The influence of noise for strong $i(x)$'s is presented in Fig 5 for the same network ($f_r = 1/5$). For $A = 0.5$ (squares), S

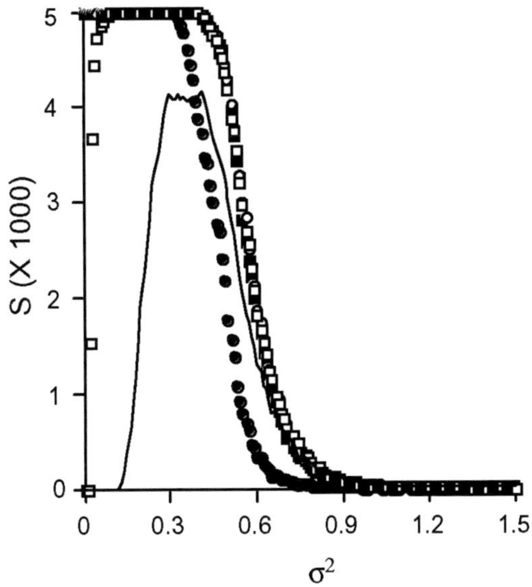

FIGURE 5. $S(\sigma^2)$ for $A = 0.5$, and $f_r = 10$: $A = 0.5$ (squares), 0.55 (circles), 0.6 (filled squares), and 0.95 (filled circles). Also shown is the $S(\sigma^2)$ for $A = 0.4$ (solid line).

increases rapidly to $S = 4900$ from $S = 0$ within: $0.01 \leq \sigma^2 \leq 0.07$. It is well-described by a $4th$-order polynomial in σ^2. For $A > 0.5$, perfect recognition ($S = 5000$) remains possible within: $0 \leq \sigma^2 \leq 0.35$. In the same σ^2-range ($A > 0.5$), the network ignores other $i(x)$'s with $f \neq 1/5$. Both are manifestations of the tolerance of trained networks to noisy strong $i(x)$'s. For $A = 0.5$, network performance eventually deteriorates with increasing σ^2 - an exponential decrease is noted for $S(\sigma^2)$ for $\sigma^2 > 0.35$.

The $S(\sigma^2)$ for $A = 0.4$, decreases to zero within the (exponential decay) bounds of the $S(\sigma^2)$'s for $A \geq 0.5$. Hence, the σ^2-range that is useful for weak-signal detection does not compromise network performance for strong $i(x)$'s - a collective behavior that is not found with individual neurons [13, 14]. The results in Fig 5 reconcile two seemingly-opposing responses of a 'mature' network to ambient noise.

Discussion

In ambient (uniform-white or Gaussian) noise, beneficial collective characteristics emerge from a network of integrate-and-fire neurons. The M-input network yields a sharper $S(\sigma^2)$ than that of a simple threshold system. Within: $5 < H \leq 35$, the same onset value and peak location range were observed for the $S(\sigma^2)$'s; where: $< S_p >= 4293 \pm 237$, and $< S_p >= 670 \pm 284$, for uniform-white and Gaussian noise, respectively. Networks with $H = 5 = M/10$, performed very poorly ($S_p = 0$).

Networks with H values within: $5 < H \leq 35$, learn quickly and generalizes perfectly. They are equally-suited for tasks involving only strong $i(x)$'s. However, their ability to benefit from noise differs because S_p decreases with H [White noise: $S_p(H=7)=4588$, $S_p(35)=3933$; Gaussian noise: $S_p(7)=1094$, $S_p(35)=340$]. Hence, ambient noise response can define the most efficient architecture for a neural network.

A network yields a faster after-peak fall-off for $S(\sigma^2)$ with white noise - the normalized $S(\sigma^2)$ decreases faster than the fifth-power of that yielded by a simple system (see Fig 1*b*). A network could also measure much smaller A's (625% sensitivity improvement is possible) and performs frequency filtering operations which are not possible with individual integrate-and-fire neurons. For a given A, the network yields a higher onset for $S(\sigma^2)$ that makes it more robust against accidental triggering due to inherent fluctuations in real physical systems.

Our work illustrates that ambient noise of the right strength ($0.1 \leq \sigma^2 \leq 0.5$ for $A = 0.4$) enhances the sensitivity of a mature network without jeopardizing its performance in strong signals. We did not observe the same collective behavior in networks of neurons with sinusoidal or linear responses.

NOISE-DRIVEN ENHANCEMENT OF SPATIAL SIGNALS

Similar to the case of Section 2, a three layer NN is used to categorize spatial signals of neural activity. During training, we feed inputs of form: $x_k = S\exp(\frac{-[k-X]^2}{2\sigma})/(\pi\sigma)^{1/2}$, i.e. Gaussian profiles with strength (S), mean (X) and variance (σ). Learning was achieved by updating the weights so that these inputs are mapped into one of the 3 desired memory states ($\mathbf{\Psi}^1, \mathbf{\Psi}^2, \mathbf{\Psi}^3$). These were chosen to be Gaussian profiles with identical strength and variance ($S = 1$, $\sigma = 2$) but varying mean ($X^1 = 7$, $X^2 = 13$, $X^3 = 19$).

Spatial information is assumed to be contained in the mean of the input state $\mathbf{x}^{t=0}$. The state to evolve to among the $\mathbf{\Psi}^i$ is therefore chosen to be the one with the closest mean, or the $\mathbf{\Psi}^i$ with smaller X^i in the special case of equal distance.

The set of training states was defined by the following 12×10^6 possible parameter combinations: $X^i - 2 \leq X^{t=0} \leq X^i + 2$ (with step size $\Delta X^{t=0} = 2.0 \times 10^{-3}$); $0.5 \leq \sigma^{t=0} \leq 3.5$ ($\Delta\sigma^{t=0} = 1.5 \times 10^{-3}$); and $S^{t=0} = 1.0$. The input states with $X^{t=0}$ equal to one of the desired means X^i will be referred to as states with "correct" mean.

After training, $\{v_{jk}\}$ are then re-wired to the input nodes as depicted in Fig. 6*b*. The resulting recurrently reconnected neural network (RNN) is now composed of $(N(1+J))$ purely thresholding TNs. The RNN is initiated with an input profile $\mathbf{x}^{t=0}$ and then evolves recurrently, with output $\mathbf{\Phi}^t$ at step t becoming the input $\mathbf{x}^{t+1}$ at step $t+1$. Note that the state of the RNN after one iteration, $\mathbf{x}^{t=1}$, is equivalent to the output of the corresponding non-recurrent network (non-RNN; Fig. 6a). Throughout the discussion, we use a network with $N = 25$ input neurons and $3N$ hidden neurons [15].

The results discussed below (i.e., noise-mediated categorization and uniformity of success rate) are obtained also in the network in which the thresholding functions f and f_H are both chosen to be $1/(1+\exp(-x))$ or $1.7159\tanh(2x/3)$ [16]. The "mixed" choice $f = 1/(1+\exp(-x))$ and $f_H = 1.7159\tanh(2x/3)$ results in fast convergence during training and directly corresponds to the standard implementation of the 2-layer perceptron [16, 17].

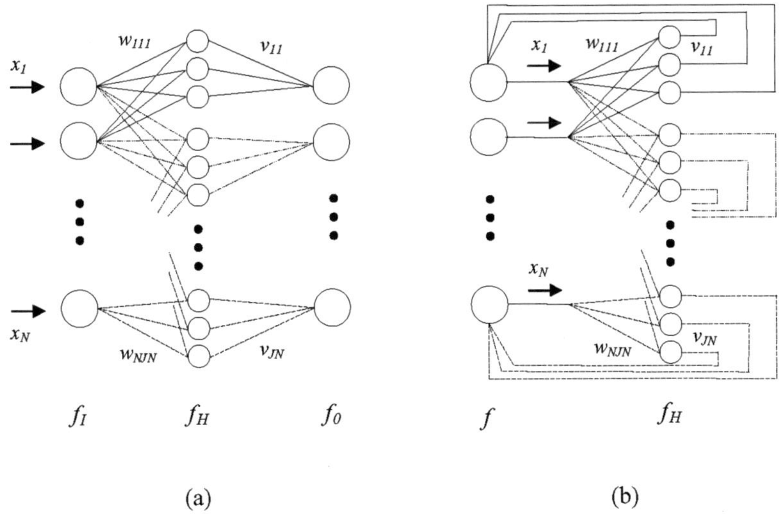

FIGURE 6. *Network architecture. a)* Three-layer fully connected feed-forward network where $f_I(x) = x, f_H(x) = 1.7159\tanh(2x/3)$, and $f_O(x) = 1/(1+\exp(-x))$. *b)* The collapsed two-layer recurent network, now with each node acting as a thresholding neuron. The thresholding function is chosen to be $f(x) = f_I\left(f_O(x)\right) = 1/(1+\exp(-x))$ for the input layer, and remains $f_H(x) = 1.7159\tanh(2x/3)$ for the hidden layer.

The similarity of $\boldsymbol{x}^t$ to the target state $\boldsymbol{\Psi}^i$ was evaluated using Linfoot's measures of structural content (C), fidelity (F), and correlation quality (Q) [18]. C provides a measure of relative sharpness of the two states, F measures their general similarity and is just the mean square error subtracted from 1, and Q quantifies the alignment of their peaks. The two states are identical when $C = F = Q = 1$. In the following, we declare that the current state of the network matches a given target state when C, F, and Q are within 5% of 1 (under such a condition $\boldsymbol{\Psi}^i$ is visually indistinguishable from $\boldsymbol{x}^t$).

To evaluate convergence to a target state in the case of the RNN, C, F, and Q are measured in 80 successive time steps ($T_{CFQ} = 80$, from $t = 921$ to $t = 1000$). We consider convergence to be complete only if the matching criterion is satisfied at each time step in this range. The extension to $T_{CFQ} > 1$ assures that the recovery measure is robust against transients (temporary trappings) in the presence of noise. For the non-RNN case, $T_{CFQ} = 1$ as convergence is evaluated at $\boldsymbol{x}^{t=1}$.

Network in the absence of noise

Following 10^5 epochs of training [20], the non-recurrent network can categorize perfectly the 12×10^6 possible inputs. Generalization by interpolation of resolutions smaller than the training resolutions is achieved perfectly within the training range. Furthermore, extrapolation to include the ranges: $0.6 < S^{t=0} < 2.0$, and $\sigma^{t=0} < 60.0$,

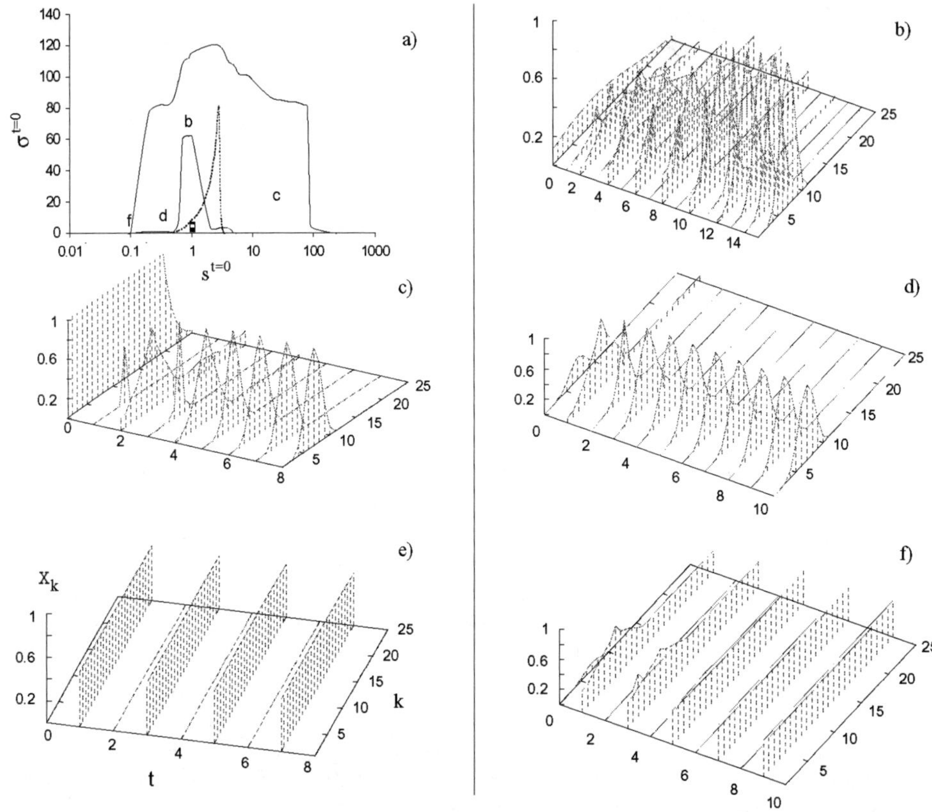

FIGURE 7. *Response of the recurrently-reconnected network to stimulus in the absence of noise.* a) The region under the solid curve represents the dynamic range where the RNN (outer solid line) and non-RNN (inner solid line) can operate perfectly for $|X^{t=0} - X^i| \leq 2$. The non-RNN network was trained to classify in the smaller region represented by the solid rectangle. Also shown is the dynamic range of the Hopfield network (dotted curve, see section III.D). Note that the scale for stimulus strength $S^{t=0}$ is logarithmic. b-f) Dynamics of the RNN starting with input states that are representative of the regions defined in Fig 7a: b) highly dispersed stimulus : $\sigma^{t=0} = 65.0, X^{t=0} = 9.0, S^{t=0} = 1.0$; c) strong stimulus : $\sigma^{t=0} = 20.0, X^{t=0} = 9.0, S^{t=0} = 50.0$; d) weak stimulus: $\sigma^{t=0} = 2.0, X^{t=0} = 7.0, S^{t=0} = 0.5$; e) no stimulus $S^{t=0} = 0.0$; f) very weak stimulus: $\sigma^{t=0} = 2.0, X^{t=0} = 7.0, S^{t=0} = 0.1$. Note that $t = 1$ corresponds to the non-RNN case.

is also attained (Fig 7a).

By adding recurrence, the dynamic range is further improved by two orders of magnitude in strength ($0.12 < S^{t=0} < 100.0$), and by a factor of two in variance ($\sigma^{t=0} < 120.0$) as shown in Fig 7a. Shown in Figs. 7b-7f are trajectories for representative stimuli from the regions in Fig. 7a in which the non-RNN fails, but the recurrently reconnected network achieves perfect categorization after time t_r. Again, note that t=1 corresponds to the non-RNN case.

Figure 7*b* shows that the state with high variance is correctly classified, i.e. converges to the state Ψ^1 with the correct mean after $t_r = 13$. The non-RNN fails to classify correctly as its output $x^{t=1}$ is dissimilar to Ψ^1. Likewise, an initially strong signal (S =50; Fig. 7c)or weak signal (S =0.5; Fig 7d) achieves recovery at $t_r = 3$ and $t_r = 5$.

When zero stimulus is presented to the RNN (Fig. 7*e*), the network reaches a limit cycle in which it alternates between $x^t = 0$ and a uniform state $x^t = 0.5$ with the constant 0.5 traceable from Eqn. 1 [21]. This is also the case for the evolution of very weak signals $S^{t=0} < 0.12$ (Fig. 7*f*), resulting in classification failure. Hence, $S^{t=0} < 0.12$ defines our set of subthreshold stimuli. It is in this region that we demonstrate the "stochastic-resonance-like" behavior of our system in the presence of noise.

Interplay of recurrence and noise

We first look at the dynamics of the RNN in the presence of pure noise (stimulus strength $S^{t=0} = 0$). Uniform white noise (UWN) of width α and zero mean was added to the state x^t at each time step, starting at $t = 0$ [22]. In a real biological network such noise can correspond e.g. to the spontaneous activity in the input neurons or to synaptic noise. The most probable response [24] of the network to pure noise can be classified as follows. For weak noise ($\alpha \leq 0.16$; Fig. 8*a*), the RNN evolution resembles the limit cycle dynamics obtained for zero noise and subthreshold stimulus. For moderate noise ($0.16 < \alpha \leq 0.29$; Fig. 8*b*), the system has a tendency to be permanently trapped in one of the memory states (Ψ^2 in case of Fig. 8*b*). For strong noise ($\alpha > 0.29$; Fig. 8*c*), x^t is trapped in the memory states Ψ^i transiently and visits all of them at random.

We now discuss three phenomena exhibited by the network categorizing in the presence of noise when recurrence is imposed: 1) Enhanced robustness to noise, 2) Uniformity of success rate throughout the dynamic range, and 3) Noise-mediated categorization.

Enhanced robustness to noise

To evaluate the performance of the network in the presence of noise, we define the success rate, S_{rate}, as the fraction of runs starting with a given stimulus that resulted in complete convergence to the correct memory state. (In the presence of noise, the Linfoot criteria are evaluated on the output x^t of the network, before noise is added.) Figure 9 shows S_{rate} for the non-RNN (Fig. 9*a*) and RNN (Fig. 9*b*) as a function of noise strength α. At signal strength $0.2 \leq S^{t=0} \leq 50$, the near-perfect performance of the RNN is robust against noise up to $\alpha \simeq 0.4$ (Fig. 9*b*). In contrast, for the non-RNN, S_{rate} decays with increasing α, and reaches nearly 0 at $\alpha = 0.4$ for $S^{t=0} < 0.5$ (Fig. 9a).

For $\alpha > 0.4$ and $0.5 < S^{t=0} < 2$, it appears at first that the non-RNN is more robust against noise than the RNN. This is no longer the case, however, if we require the non-RNN to satisfy the Linfoot criteria for 80 successive runs with identical stimulus (i.e. a criterion comparable to the one used for the RNN). At $\alpha \approx 0.11$, S_{rate} then abruptly becomes zero (figure not shown) for the non-RNN.

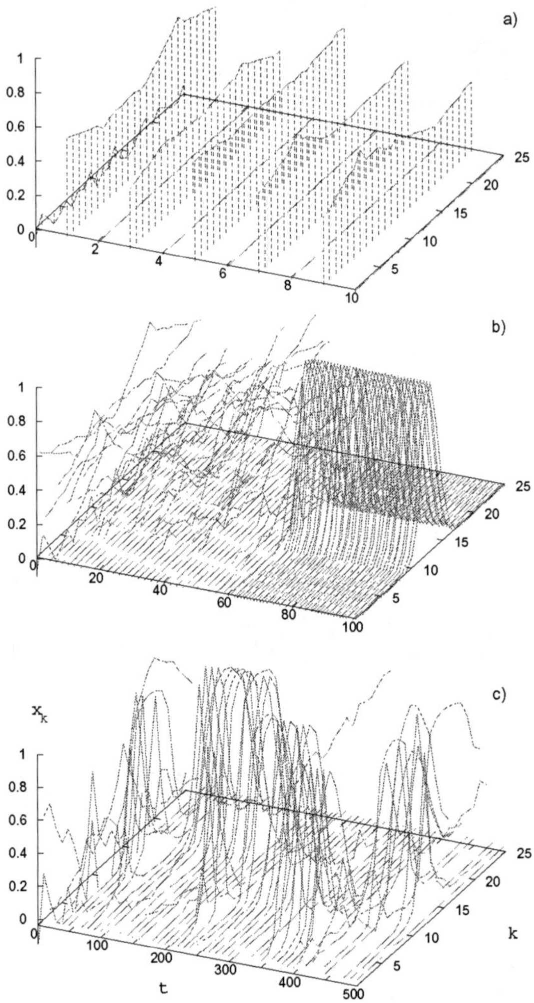

FIGURE 8. *Typical recurrently-reconnected-network dynamics for zero stimulus* and a) weak noise ($\alpha = 0.14$); b) moderate noise ($\alpha = 0.22$), c) strong noise ($\alpha = 0.50$). The output $\boldsymbol{x}^t$ of the network (before noise is added) is shown for $t \geq 1$; for $t = 0$, $\boldsymbol{x}^0$ plus UWN is shown.

Without recurrence, even in the absence of noise, the non-RNN fails when $S^{t=0} < 0.6$ or $S^{t=0} > 2.0$. The non-RNN therefore has a dynamic range (DR) of less than one decade, with low robustness against noise. In contrast, RNN possesses a DR of over two decades in $S^{t=0}$ (0.12 to 85.0) and is robust up to noise strength $\alpha \approx 0.4$)independent of $S^{t=0}$.

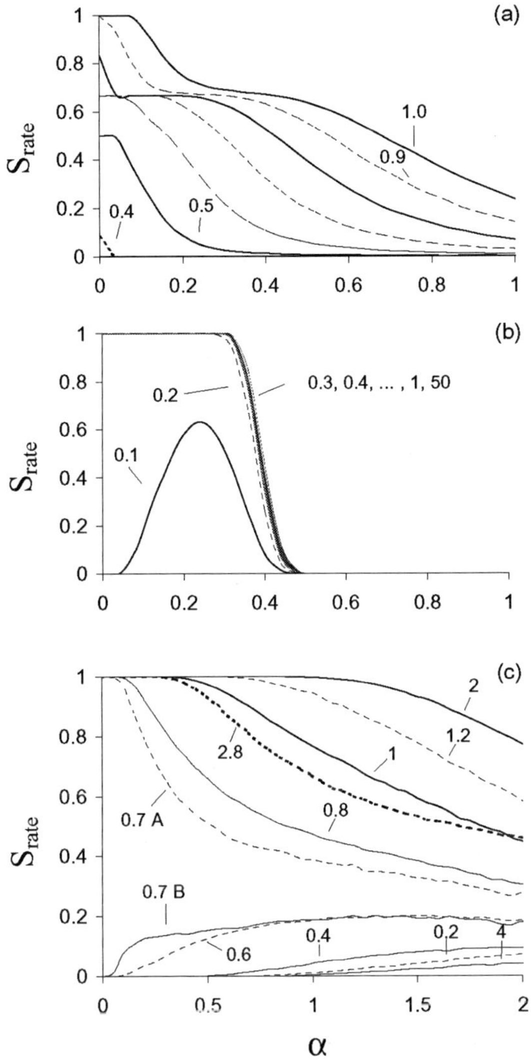

FIGURE 9. *Success rate of a) non-recurrent, b) recurrently-reconnected, and c) Hopfield networks in the presence of noise.* Average S_{rate} is shown as a function of noise strength α at the indicated stimuli strengths $S^{t=0}$. In a) and b), S_{rate} is averaged over stimuli with $\sigma^{t=0} = 2$ and $X^{t=0}$ in the range $X^i - 2 \leq X^{t=0} \leq X^i + 2$ (resolution of $X^{t=0} = 1.0$), with 50 runs performed for each $X^{t=0}$. In c), all curves except 0.7A and 0.7B were averaged over stimuli with $\sigma^{t=0} = 4$ and with $X^{t=0}$ as in a) and b). The curve 0.7A shows S_{rate} for stimuli with strength $S = 0.7$ and "correct" mean $X^{t=0} = X^i$, while the curve 0.7B shows the average over $X^{t=0} = X^i \pm 2$ and $X^i \pm 1$. 1000 runs were performed for each stimulus in c).

Uniformity of success rate throughout dynamic range

The success rate of the RNN is approximately uniform throughout its dynamic range: $S_{rate} \simeq 1$ for $\alpha < 0.4$ and $S_{rate} \simeq 0$ for $\alpha > 0.5$, independently of signal strength $S^{t=0}$ (Fig. 9*b*). To verify that this result is not a consequence of the strict condition we imposed on S_{rate}, we lowered both T_{CFQ} (down to $T_{CFQ} = 1$; only evaluated at $t = 1000$) and Linfoot's tolerance (up to 20% variation of C, F, Q from 1.0). We obtained curves that are similar to Fig 9*b* except that at higher noise strengths ($\alpha > 0.5$), S_{rate} saturates at a non-zero value (0.2 for $T_{CFQ} = 1$ and 20% tolerance). The non-zero sucess rate at high noise strength originates from the transient (typically less than 10 iterations) trapping of the system in one of the incorrect memory states $\boldsymbol{\Psi}^i$ (Fig. 8*e*) — the reason why we set T_{CFQ} to a relatively high value ($= 80$) in Fig. 9*b*. The corresponding $S_{rate} \approx 0.2$ is lower than $1/3$ as in the presence of noise of high strength, the network state is not always close to one of the $\boldsymbol{\Psi}^i$.

Noise-mediated categorization

In the absence of noise, the RNN fails to respond to stimuli of strength $S^{t=0} < 0.12$ (i.e., it evolves as if no stimulus is presented, Fig. 7e and 7f). When noise is added, however, such subthreshold stimuli can be classified correctly. This is clearly seen for $S^{t=0} = 0.1$ in Fig. 9*b*, and $S^{t=0} < 0.12$ in Fig. 10*a*. The shape of the $S_{rate}(\alpha)$ curves is reminiscent of stochastic resonance [11, 23, 1]: the highest S_{rate} is obtained at a non-zero optimal noise strength, α_{opt}. The average S_{rate} for signals of strength $S^{t=0} = 0.1$ reaches 60% at $\alpha_{opt} \approx 0.24$. Below $S^{t=0} = 0.12$, the optimal S_{rate} decays linearly with $S^{t=0}$, reaching 30% at $S^{t=0} = 0$. The 30% $\approx 1/3$ level reflects the long-term trapping of the network in a randomly selected memory state (Fig. 8b) when α falls into the moderate noise range. $S_{rate} = 1/3$ therefore represents the baseline to which the gain in S_{rate} should be compared. Still, adding noise at strength α_{opt} increases S_{rate} by up to 100% (from 30% to 60%).

The addition of noise also improves the classification of signals (of arbitrary strength) that were not included in the training set of the NN ("noise-mediated extrapolation"). This is apparent in Fig. 10*b* for signals with $X^{t=0} \simeq 10$ or $X^{t=0} \simeq 16$ (in the region of strong overlap between two of the states $\boldsymbol{\Psi}^i$) and $X^{t=0} < 4$ or $X^{t=0} > 22$ (near the system boundary).

The RNN viewed as a dynamical system

The recurrently-reconnected network implements a discrete dynamical system that maps $\boldsymbol{x}^t$ into $\boldsymbol{x}^{t+1}$ through: $\boldsymbol{x}^{t+1} = \Gamma(\boldsymbol{x}^t)$, where the mapping function Γ is given by eqn (1).

By construction, the three target states $\boldsymbol{\Psi}^i$ are fixed points of the map. Note that the state $\boldsymbol{x} = 0$ is not a fixed point. To evaluate the stability of $\boldsymbol{\Psi}^i$, we examine the

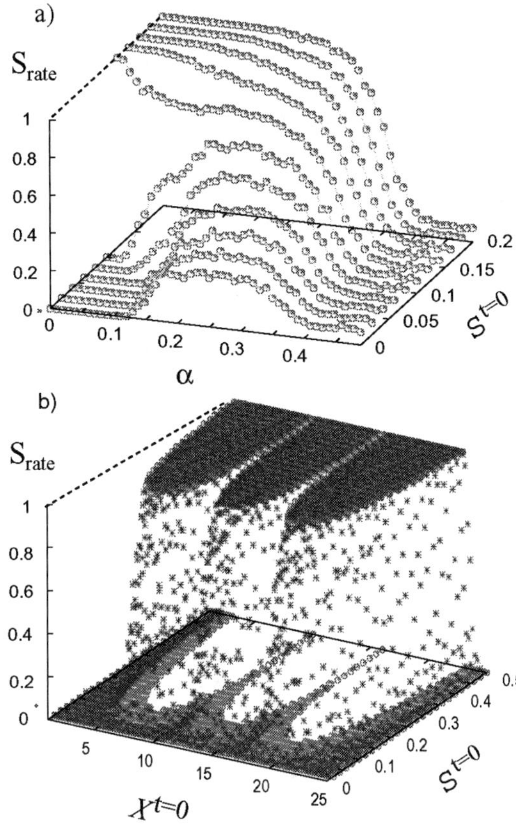

FIGURE 10. *Noise -mediated categorization of weak stimuli by the recurrently-reconnected network.* *a*) Success rate of the RNN averaged over stimuli with $X^i - 3 < X^{t=0} < X^i + 3$. b) Success rate of the RNN in the absence ($\circ$) and presence ($*$) of noise with strength $\alpha = 0.24$. In all cases $\sigma^{t=0} = 2.0$. Linfoot's criteria are evaluated with a $T_{CFQ} = 10$ from $t = 991$ to $t = 1000$ and averaging is done over 50 runs for each stimulus.

eigenvalues $\boldsymbol{\lambda}$ of the Jacobian matrix $\boldsymbol{\Omega}$ of the system [25]. Fig. 11*a* shows that the largest modulus of $\boldsymbol{\lambda}$ is ≈ 0.21; as $|\lambda_k| < 1$ for all k, $\boldsymbol{\Psi}^i$ are stable fixed points [26].

The trajectories $\boldsymbol{x}^t$ for the noiseless case are visualized in Fig. 11*b* , where the three axes (d_1, d_2, d_3) are the vector projections of $\boldsymbol{x}^t$ on the three target states $\boldsymbol{\Psi}^i$, in the units of $|\boldsymbol{\Psi}^i|$ (i.e. $d_i = (\boldsymbol{\Psi}^i \cdot \boldsymbol{x}^t)/|\boldsymbol{\Psi}^i|^2$). The trajectories form four clusters, three of which converge to one of $\boldsymbol{\Psi}^i$, and the fourth cluster consists of a thin region (with thickness $\Delta d \approx 0.24$) oriented along the diagonal directions (see Fig. 11*b*). In the latter region, $\boldsymbol{x}^t$ asymptotically alternates between $(d1, d2, d3) = (0.00, 0.00, 0.00)$ and $(0.68, 0.69, 0.68)$. The attractor of this region, which includes the zero stimulus trajectory, is thus a limit cycle of period two.

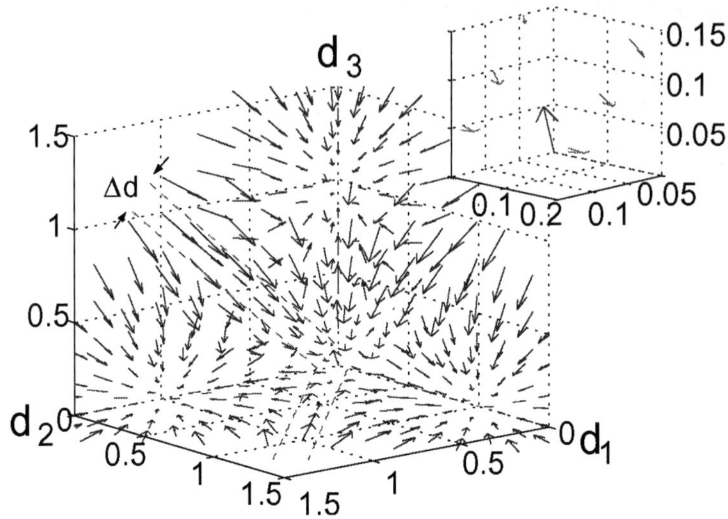

FIGURE 11. *Flow diagram of the recurrently-reconnected network.* Shown are the projections d_i of the state $\boldsymbol{x}^t$ on the memory states $\boldsymbol{\Psi}^i$. Included are trajectories for stimuli from the set $0.0 \leq S^{t=0} \leq 2.0$; $1.0 \leq X^{t=0} \leq 25.0$; $1.0 \leq \sigma^{t=0} \leq 25.0$. Inset: The "velocity vector" $\boldsymbol{d}^{t+1} - \boldsymbol{d}^t$ with vector origin defined by $\boldsymbol{d}^t$. The velocity at $\boldsymbol{x}^{t=0} = 0$ has a strong magnitude and moves the system to the flat state $\boldsymbol{x}^t$ with $(d_1 = 0.68, d_2 = 0.69, d_3 = 0.68)$. The magnitude of the velocities is scaled to 0.2 of its original strength.

We are now in a position to understand the results of Sect. III B in terms of the flow represented in Fig. 11b. The weaker the stimulus strength $S^{t=0}$, the higher is the percentage of stimuli of this strength that become trapped in the limit cycle attractor, rather than converging to one of the desired fixed points $\boldsymbol{\Psi}^i$; with no noise, all states with $S^{t=0} < 0.1$ become trapped. In the presence of noise, the trajectory can switch in between the basins of attraction of the zero-noise system. When the noise strength α becomes comparable to the thickness $\Delta d \approx 0.24$, the trajectories starting in the undesired basin are likely to escape from it. If noise is too strong ($\alpha > 0.29$), the trajectory is unlikely to become permanently trapped in the vicinity of one of the target states, and instead hops randomly between the corresponding basins of attraction (see also Fig. 8c). The existence of the optimal value of noise strength α_{opt}, is then seen to be a compromise maximizing the chance of getting out of the undesired basin and subsequently not escaping the desired basin.

Comparison to a Hopfield network

The arguments of the previous section are quite general, and raise the question of whether the effects discussed by us are a generic feature of recurrent networks. In this section, we describe a simple associative memory network that fails to exhibit the enhancement of performance by noise which we demonstrated in the two-layer recurrently reconnected network (RNN). As described in Appendix B, we constructed a

Hopfield network with three target states ${}^{H}\Psi^q$ of means $X^1 = 7, X^2 = 13$ and $X^3 = 19$. The states are defined by $\{{}^{H}\Psi^q_k = 1\}$ for $k = X^q \pm 2, X^q \pm 1, X^q$ and $\{{}^{H}\Psi^q_k = -1\}$ for all other k (k being the index of each neuron). We then evaluated the performance of the network for the set of input states defined by

$$x_k = \frac{2S\exp(\frac{-[k-X]^2}{2\sigma})}{(\sigma/4)^{1/2}} - 1 \tag{1}$$

Note that when the sgn function is applied to the input states with $S = 1$, $\sigma = 4$ and $X = X^q$, one obtains the target states ${}^{H}\Psi^q$. Other input states are considered to be correctly classified if the network evolves to the target state with closest mean.

We first discuss the dynamics in the absence of noise. The input state $\{x_k = -1\}$ (corresponding to $S^{t=0} = 0$) represents the zero stimulus and does not evolve, as $\mathbf{x}^t = -1$ is a fixed point of the map. The set of subthreshold stimuli is defined as those that converge to the fixed point $\{x_k = -1\}$ after one iteration. The dynamic range of the network (within which perfect classification is achieved) is indicated in Fig. 7a. Note that the dynamic range in $S^{t=0}$ is two orders of magnitude narrower compared to the range of the two-layer RNN.

The success rate S_{rate} of the Hopfield network in the presence of noise is plotted in Fig. 9c. In contrast to the RNN case (Fig. 9b), the success rate is not uniform as a function of signal strength. Within most of its narrow dynamic range, the Hopfield network is more robust than the RNN.

For large enough noise strength α, the success rate of the Hopfield network does become non-zero (Fig. 9c) even for subthreshold stimuli ($S^{t=0} \leq 0.6$). However, S_{rate} never exceeds $1/3$, i.e. the success rate that corresponds to classification performed at random. Therefore the implemented Hopfield network does not show noise-mediated categorization.

This illustrates that the conditions necessary for noise-mediated detection of subthreshold stimuli will not always be met in recurrent neural networks. The failure of the Hopfield network to benefit from noise can be understood using arguments similar to those of Section III.C. Compared to the RNN case, the size of the basin of attraction for the zero-stimulus fixed point $\{x_k = -1\}$ is larger, both in absolute terms and relative to the size of basins of attraction for the memory states ${}^{H}\Psi^q$ [27]. In addition, the dynamics is strongly influenced by the spurious fixed points $\mathrm{sgn}(\pm{}^{H}\Psi^1 \pm{}^{H}\Psi^2 \pm{}^{H}\Psi^3)$. When designing a neural network capable of noise-mediated categorization, particular care should be given to the minimization of the basins of attraction of unwanted attractors.

Related Ideas

We first briefly discuss our results in the context of biological neural systems. Consider a recurrent neural network that processes spatially encoded sensory information, for example the olfactory bulb network [28]. The input to the network is generated by sensory neurons at the periphery (the olfactory epithelium). Even in the absence of sensory stimulus, many sensory neurons have high levels of spontaneous activity (≈ 100

spikes /sec in typical vertebrate olfactory sensory neurons [29]). Following the presentation of a brief sensory stimulus, the sensory neurons provide a transient, spatially encoded input to the network [30]. If the interval between successive presentations of stimuli allows sufficient time for the network to evolve recurrently [31], the output from the network will show the noise-mediated features discussed above. Spontaneous activity in the sensory neurons can in this context be viewed as a necessary part of a biological mechanism that optimizes the detection of weak stimuli by the system.

Identification and discrimination of odors can be viewed as a categorization task performed by the olfactory system. Psychophysical studies show that the perceived quality of an odor is typically invariant to odor concentration over two orders of magnitude [32]. Recently, Brody and Hopfield proposed a phenomenological model based on Many-are-Equal operations, designed in part to explain this invariance [33]. Our results show that in the presence of noise such a feature can emerge also by allowing simple recurrence.

SUMMARY

In summary, we presented a detailed case study designed to investigate the effect of noise in neural networks. We showed that noise of proper variance introduced to the receiving neurons allows instantaneous improvement of temporal signal recovery of NN. Also in the presence of noise, the introduction of recurrence extends and homogenizes the operating range of NN trained to categorize spatial patterns of neural activity. We also gave an example of a simpler recurrent network of the Hopfield type where noise-mediated categorization fails. We gave an intuitive explanation of the difference between these two cases based on the fixed point structure of the two systems.

APPENDIX A: TRAINING PROCEDURE FOR THE RECURRENTLY RECONNECTED NETWORK

The synaptic weights $\{w_{ijk}\}$ and $\{v_{jk}\}$ were initialized randomly within $[-1/\chi, 1/\chi]$ where χ is the number of free parameters connected to the weights ($\chi = JN$ and $\chi = J$, respectively for $\{w_{ijk}\}$ and $\{v_{jk}\}$) [17, 34]. The activation functions for the input, hidden and output nodes are $f_I(x) = x, f_H(x) = 1.7159\tanh(2.0x/3.0)$, and $f_O(x) = 1/1+\exp(-x)$, respectively. The threshold function f_H is chosen so as to allow faster convergence and better memory representation [17, 34, 35]. The linear function f_I in this (training) stage is chosen so that the network can be wrapped in the next (recurrent) stage as depicted in Fig. 6*b*, hence, eventually producing a network with $N(1+J)$ purely TNs.

Fixation of the desired memory state $\boldsymbol{\Psi}$ for an initial profile $\boldsymbol{x}^{t=0}$ is attained through gradient descent by updating the weights according to:

$$w_{ijk}^{t+1} = w_{ijk}^{t} + \eta \frac{\partial E_k}{\partial w_{ijk}} + \gamma \Delta w_{ijk}^{t-1} \quad (2)$$

where $E_k = (\Psi_k - \Phi_k)^2/2$ is the cost function, and $\gamma\Delta w_{ijk}^{t-1} = \gamma(w_{ijk}^{t-2} - w_{ijk}^{t-1})$ is the momentum term [17, 34] . The learning (η) and momentum (γ) rates are adaptively varied within ($10^{-5} \leq (\eta,\gamma) \leq 10^{-1}$) to account for the variation of the error surface along different regions of weight dimension [36]. These rules enhance the minimization of E_k, preventing unnecessary oscillations in the search space [17, 34, 35, 36]. A similar update rule is implemented for v_{ij}.

APPENDIX B: IMPLEMENTATION OF THE HOPFIELD NETWORK

The one-layer, recurrent, associative memory network of the Hopfield type was designed to perform a classification task similar to that for which we trained the network of Appendix A. The desired memory states ${}^{\mathrm{H}}\boldsymbol{\Psi}^q$ were chosen to be the discretized version of the Gaussian profiles that were used as memory states of the RNN:

$$ {}^{\mathrm{H}}\Psi_k^q = \mathrm{sgn}\Big(\frac{2S\exp(\frac{-[k-X^q]^2}{2\sigma})}{(\sigma/4)^{1/2}} - 1\Big) \tag{3}$$

where $\sigma = 4$, and $S = 1$. The means (X^q) that characterize the three memory states are $X^1 = 7, X^2 = 13$ and $X^3 = 19$ — in exact correspondence to those in Appendix A.

The synaptic weight (w_{ij}) that connects the ith neuron to the jth neuron is fixed using the Hebbian rule [17, 35]

$$ w_{ji} = w_{ji} = \begin{cases} \frac{1}{N}\sum_{q=1}^{3}({}^{\mathrm{H}}\Psi_i^q)({}^{\mathrm{H}}\Psi_j^q) & j \neq i \\ 0 & j = i \end{cases} \tag{4}$$

The state of the network $\boldsymbol{x}^t$ evolves recurrently according to the map

$$ x_j^{t+1} = \mathrm{sgn}\Big(\sum_{i=1}^{25} w_{ji}x_i^t\Big), \qquad j = 1,2,...,25 \tag{5}$$

The update of $\boldsymbol{x}^t$ is done asynchronously, i.e. the element j is chosen randomly and one at a time.

REFERENCES

1. C. Monterola and C. Saloma, Phys. Rev. Lett. **89**, 188102(2002).
2. C. Monterola and M. Zapotocky, Phys. Rev. E **71**, 036134(2005).
3. L. Gammaitoni, P. Hänggi, P. Jung, and F. Marchesoni, Rev. Mod. Phys. **70**, 223-287(1998).
4. A. Longtin, A. Bulsara, and F. Moss, Phys. Rev. Lett.**67**, 656 (1991).
5. M.C. Teich, S. Khanna and P. Guinney, J. Stat. Phys. **70**,257 (1993).
6. D. Russel, L. Wilkens and F. Moss, Nature **402**, 291(1997).
7. J. Levin and J.P. Miller, Nature **380**, 165 (1996).
8. E. Simonotto, et al, Phys. Rev. Lett. **78**, 1186 (1997).
9. I. Hidaka, D. Nozaki and Y. Yamamoto, Phys. Rev. Lett. **85**, 3740 (2000).
10. J. Proakis and D. Manolakis, Digital Signal Processing: Principles, Algorithms and Applications (Prentice-Hall, New York, 1995).

11. L. Gammaitoni, Phys. Rev. E. **52**, 4691 (1995).
12. N. Mohanty, Signal Processing: Signals, Filtering, and Detection (Van Nostrand Reinhold, New York, 1987).
13. J. Feng and B. Tirozzi, Phys. Rev. E **61**, 4207 (2000).
14. N. Brunel, et al, Phys. Rev. Lett. **86**, 2186 (2001).
15. Results analogous to those discussed for $N = 25$ were obtained for $N \leq 100$ ($N = 4, 9, 16, 50, 100$) and an arbitrary number of memory states below $N/2$. We chose $J = 3$ to obtain a network of high capacity per neuron. (The capacity of the 3-layer feedforward network is $C = \frac{16}{\pi-2}(\ln(J))^{1/2}N$
16. Results for the network with $f = 1/(1+\exp(-x))$ and $f_H = 1/(1+\exp(-x))$ are given at http://www.mpipks-dresden.mpg.de/ chris. In this case, the period-two limit cycle of the RNN (Sec. III.A) is replaced by a fixed point $\boldsymbol{x} \simeq 0$.
17. S. Haykin, *Neural Networks: A Comprehensive Foundation, 2nd Ed.* (Prentice-Hall, New York, 1999).
18. The expressions for Linfoot's criterion are: $C = \langle \boldsymbol{x}^t \rangle^2 / \langle \boldsymbol{\Psi}^i \rangle^2$; $F = 1 - \langle \boldsymbol{\Psi}^i - \boldsymbol{x}^t \rangle^2 / \langle \boldsymbol{\Psi}^i \rangle^2$ and $Q = \langle |\boldsymbol{\Psi}^i||\boldsymbol{x}^t| \rangle / \langle \boldsymbol{\Psi}^i \rangle^2$ where $\langle \quad \rangle$ designates averaging over the state (note that $F = 2Q - C$) [19].
19. F. Huck *et al.*, J. Opt. Soc. of Am. **2**, 1644(1985).
20. During each training epoch, the full set of 12×10^6 stimuli was presented in random order. The NN was implemented in c++ and compiled using gcc; the run time for training was 40 minutes using a DEC ALPHA computer with 1 GHz processor speed. We note that the results are independent of the initial weights. Even though the set of the final values of the weights are different for each weight initialization, results consistent with those described in section III (i.e. oscillation between 0 and 0.5, noise-enhanced categorization, etc) are obtained for each of the 50 initializations we tested. A representative set of the obtained weights $\{w_{ijk}\}$ and $\{v_{jk}\}$ are listed at `http://www.mpipks-dresden.mpg.de/~chris`.
21. The transition $0 \to 0.5$ follows immediately from Eqn. 1: $Y_{jk} = f_H(0) = 1.7159 * \tanh(0) = 0$; $\Phi_k = f_O(0) = 1/(1+\exp(-0)) = 0.5$. The transition $0.5 \to 0$ occurs as the sum of the inputs to each unit in the hidden layer is a (large) negative number when the state $\boldsymbol{x}^t$ at the previous time step was uniform. This is a property of the evolved weights.
22. Noise was generated using the $MT19937$ generator by Matsumoto & Nishimura (prime period $\approx 10^{6000}$) taken from the GNU scientific library with seed randomized by the CPU clock. Results analogous to those reported for UWN noise were also obtained for Gaussian-distributed noise.
23. G. Wenning and K. Obermayer, Phys. Rev. Lett. **90**, 120602 (2003); S. Laughlin and T. Sejnowski, Science **301**, 1870(2003); T. Hoch, G. Wenning, and C. Obermayer, Phys. Rev. E **68**, 011911(2003); B. Kosko and S. Mitaim, Neural Networks **16**, 755(2003); M. Chacron, A. Longtin and L. Maler, Network: Comput. Neural Syst. **14**, 803(2003); I. Goychuk, and P. Hänggi, Phys. Rev. E **69**, 021104 (2004).
24. The behavior described in the main text was obtained in 93% of our runs for weak noise ($\alpha < 0.16 (\Delta\alpha = 0.01)$), 81% of runs for moderate noise ($0.16 \leq \alpha < 0.29 (\Delta\alpha = 0.01)$), and 87% of runs for strong noise ($0.29 \leq \alpha \leq 1.0 (\Delta\alpha = 0.05)$). 10 runs were performed at each noise strength.
25. The explicit expression for λ_k and the corresponding eigenvectors (in a Mathematica 5.0 notepad) can be viewed at `http://www.nip.upd.edu.ph/ipl/mem/homepages/chris/`.
26. H. Tong, *Non-linear time series: a dynamical system approach* (Clarendon Press Oxford, 1990).
27. We traced the evolution of all the 2^{25} possible combinations for the 25 neurons each with a state of either -1 or 1. The fixed points where the network will eventually settle were distributed as follows: ${}^{H}\Psi^1 = 11\%$, ${}^{H}\Psi^2 = 13\%$, ${}^{H}\Psi^3 = 11\%$, $\{x_k = -1\} = 16\%$, $\{x_k = 1\} = 13\%$, $-{}^{H}\Psi^1 = 9\%$, $-{}^{H}\Psi^2 = 11\%$, $-{}^{H}\Psi^3 = 9\%$, and $\mathrm{sgn}(\pm{}^{H}\Psi^1 \pm {}^{H}\Psi^2 \pm {}^{H}\Psi^3) = 5\%$. In the last case we exclude the fixed points when the three signs are either all $(+)$ or all $(-)$ because such are identical to $\{x_k = -1\}$ and $\{x_k = 1\}$, respectively. Only about 2% of 2^{25} have settled to fixed points other than the states listed above.
28. K. Mori , H. Nagao, and Y.F. Sasaki, Network **9**, R79-102(1998).
29. J.P. Rospars *et al.*, Brain Res. **662**, 31-44 (1994).
30. This will also be the case when the sensory stimulus is extended in time and the sensory neurons rapidly adapt.
31. While we evaluated the success rate of recovery of weak signals in Fig. 10b after 1000 iterations, in most cases less than 20 iterations are required to converge to the correct memory state.
32. D. Krone *et al.*, Phytother. Res. **15**, 135(2001).

33. C. Brody and J.J. Hopfield, Neuron **37**, 843(2003).
34. T. Watkin and A. Rau, Rev. Mod. Phys. **65**, 499(1993).
35. J. Hertz, A. Krogh, and R. Palmer, *Introduction to the theory of neural computation*(Addison-Wesley Publishing Company, 1991).
36. Z. Lou, Neural Computation **3**, 226(1991); C. Monterola *et al.*, J. Forecast. **21**, 435(2002).

White Noise Path Integrals in Stochastic Neurodynamics

M. Victoria Carpio-Bernido and Christopher C. Bernido
Research Center for Theoretical Physics
Central Visayan Institute Foundation
Jagna, Bohol 6308, Philippines

Abstract. The white noise path integral approach is used in stochastic modeling of neural activity, where the primary dynamical variables are the relative membrane potentials, while information on transmembrane ionic currents is contained in the drift coefficient. The white noise path integral allows a natural framework and can be evaluated explicitly to yield a closed form for the conditional probability density.

Keywords: Neurons, Path integral, White noise analysis
PACS: 87.10.Mn, 87.19.ll, 31.15.xk

INTRODUCTION

Recent experimental results in neuroscience indicate that major factors affecting the firing of neurons are: the location, geometry, synaptic molecular dynamics and chemical composition of the neuron, and transneuronal ion currents. Recall that there are about 10^{11} neurons in the entire human brain [1]. These are of different shapes and types, and are generally arranged in layers and clusters or constellations, the geometry of which may be related to the functions of each neuronal cluster. For example, while there are clusters arranged in lattice configuration, there are others appearing as almost flat two-dimensional plates. Still other clusters seem randomly arranged throughout a cortical layer. Furthermore, each neuron has a system of dendrites and axons with synapses linking it to around 10^4 other neurons. Intercellular communication through synapses is then initiated by the transmission of electrical signals generated when a neuron 'fires', or releases ionic currents to revert to equilibrium relative membrane potentials. Experimental results for spatial mapping of such neuronal activity during performance of particular tasks show location dependence: grammar processing involves more frontal regions of the left hemisphere of the brain, while semantic processing involves the posterior lateral regions of both left and right hemispheres. Also of special interest for pedagogical applications [2] is the Triple Code Model. Visual perception of a number, say 5, involves the fusiform gyrus, hearing the number spoken activates the perisylvian area, and understanding that five is greater than one involves the interparietal lobes. Finally, neurons are dynamic biological powerhouses with the ability to grow new synapses, or retract dysfunctional ones, and to synthesize proteins, in the soma or even at synaptic sites, on demand [3], depending on identified needs and tasks.

Finer spatio-temporal mapping of the brain, its structure and functions, is being made possible by continuing advances in medical imaging such as enhanced postiron emission tomography (PET) scans and functional magnetic resonance imaging (fMRI). Another development, optical topography, allows subjects to move, unlike the PET where the subject should be immobile for 40 minutes or more for sharper image resolution. The technique developed by Koizumi et al. [4] targets higher-order brain function analysis by trans-cranial dynamic near-infrared spectroscopy imaging. On the other hand, electro- and magneto-encephalography with superconducting quantum interference devices (SQUIDS) allow high resolution temporal mapping of brain activity (of the order of 10^{-3}s). Even with advances in mapping the brain and its functions, however, much remains to be understood because, aside from the complexity of its structure, huge numbers of variables are involved in carrying out functions and tasks. Much work in different areas has been

CP1021, *Stochastic and Quantum Dynamics of Biomolecular Systems,*
Proceedings of the 5th Jagna International Workshop, edited by C. C. Bernido and M. V. Carpio-Bernido

done and includes theoretical models accommodating various features of the brain. A popular approach that itself has become a separate modeling tool for different fields of study is the use of neural networks. Each neuron is treated as a simple computer and the brain as a large composite computer [5]. There are limitations to this approach, notably, unlike two-state computer bits that are either "on" or "off", a single neuron can be activated through a continuum of states, and links with a large number of other neurons activated through their own spectrum of neuronal states. Another model views the brain as a nonlinear nonequilibrium open system with infinite degrees of freedom. Macroscopic patterns in aggregates of neurons, instead of the activity of single neurons, are considered. The evolution and dynamics of the patterns are then interpreted as cognitive flows. Such a picture treats the brain as a pattern-forming, self-organized, dynamical complex system [6].

Considering the complexity and breadth in scope of brain studies, it remains important to investigate different models, approaches and modern mathematical tools for possible insights that could lead to a deeper understanding of how the brain works. Indeed, models that are explanatory, quantitative, and predictive could have tremendous implications in the health sciences, education and learning, and the dynamics of social systems and communities. It is with this motivation that we present initial studies on a mathematical framework that may be rich enough in structure to accommodate different features of the brain: geometry and global topology, structure, large-N dimensional configuration spaces, and dynamical fluctuations typical of biological systems. We start with the parametrization idea of neuron firing introduced by K. Yasue et al. [7], identifying the dynamical random variables as the relative membrane potentials and the currents. However, we employ the path integral method to investigate the stochastic dynamics of the system. The path integral, or sum-over-all-histories approach, is a most natural framework for treating infinite-dimensional systems. This has been recognized earlier, for example, by M. Turvey, [8] who suggested the use of the path integral for modeling how the cerebellum selects and shapes a control signal from background neuronal excitation. On the other hand, the path integral in the spin coherent state representation has been used in a master equation for neural networks by Ohira and Cowan [9]. Moreover, the path integral also has the added advantage of being able to deal with nontrivial topological features [10] that may arise in modelling brain functions.

In this paper, we adopt a different use of the Feynman path integral to model features of brain processes. First of all, the path integral is interpreted in the sense of Hida and Streit, [11] i.e., as an average over a Gaussian white noise measure of well-defined distributions on the infinite dimensional coordinate system formed by the random white noise variables, $\omega(t) = dB/dt,\ t \in \mathbf{R}$. Here, B denotes the usual Brownian motion. Evaluation of the path integral yields the probability density function describing the stochastic variation in time of the relative membrane potential of the neuron.

In the next section, we shall briefly describe the white noise path parametrization approach of Hida and Streit. This is followed by a simple modeling of neuronal activity for which a probability density function for the relevant dynamical variables is obtained. Fluctuations and temporal variation in ionic currents through the neuronal membrane are then contained in a modulating function naturally appearing in the white noise path integral. The counterpart of this function in the usual stochastic differential equation approach (Langevin type) is the drift term. It also contains information about the other neurons linked to the neuron being considered.

PATH INTEGRAL MODELING OF NEURAL ACTIVITY

We designate each specific physical location of a neuron in the brain as a point $\mathbf{x} \in \mathbf{R}^3$ in the usual three-dimensional space. For example, in spherical coordinates, if we take the origin to be the geometric center of the the brain, then $\mathbf{x} = \{r, \theta, \varphi\}, 0 \leq r < r_b$, (with r_b indicating the radial extent of the brain), $-\pi \leq \theta < \pi, 0 \leq \varphi < 2\pi$. The location of the neuron also specifies the origin of the impulse it generates. To locate the three-dimensional physical position of N ($N \approx 10^{10}$) neurons, we take $\mathbf{x}_i$, where in spherical coordinates, $\mathbf{x}_i = \{r, \theta, \varphi\}_i$, $i = 1, 2, ..., N$. Clusters of neurons may form a domain, Ω. Neurons at $\mathbf{x}_i \in \Omega_a$ may be responsible for hearing, while those at $\mathbf{x}_j \in \Omega_b$, with $\Omega_a, \Omega_b \in \mathbf{R}^3$ are for sight, and so on. Furthermore, the nature of the impulse produced by any neuron depends on the stochastic temporal variation of the potential difference between intracellular and external ionic concentrations. The interior of a neuron is separated from the outside by a 5-nm-thick neuronal membrane that serves as a barrier preventing the inflow of certain substances in the extracellular fluid. There is, however, a diffusive movement of charged ions across the neuronal membrane, and an electrical field arising from voltage differences across the membrane. In the so-called resting neuron, the voltage difference arises from the cytosol along the inner surface of the membrane negatively charged compared with the outside. We can then designate the voltage across the neuronal membrane at any given time t as $V_m(t)$. This is referred to as the membrane potential. Whenever the membrane potential crosses a threshold voltage $V_T(t)$, the neuron fires and sends an impulse that propagates along its axon. To characterize the firing of neurons we can then consider a variable $v(t)$ which is the difference of two voltages:

$$v(t) = V_m(t) - V_T(t) \ . \tag{1}$$

This is called the relative membrane potential [7]. A neuron fires and is active when, $v(t) > 0$, and is quiescent when $v(t) < 0$. With this, for each neuron at $\mathbf{x}_i$ there is also associated a variable $v_i(t)$. In reference [7], stochastic fluctuations are included in the diffusion process, so that the random variable $v_i(t)$ obeys the stochastic differential equation,

$$dv_i(t) = A_i(v,t)\ dt + dw(t)\ , \tag{2}$$

where $dw(t)$ gives the Gaussian noise term.

The activity of the ith neuron and its interaction with other neurons can be described as $\eta_i = \eta_i(\mathbf{x}_i, v_i(t))$. However, in this paper, we focus only on variations in the relative membrane potentials, $v_i(t)$. In the simple case of a single neuron characterized by $v(t)$, Eq.(1), the stochastic variation of the random variable $v(t)$ between the initial value $v(0) = v_0$, and the final value at $v(t) = v$ can be characterized by the conditional probability density $p(v, t \mid v_0, t_0)$. This density function satisfies the Fokker-Planck equation corresponding to the stochastic differential equation, Eq.(2). We have

$$\begin{aligned} \frac{\partial}{\partial t} p(v,t \mid v_0,t_0) &= -\frac{\partial}{\partial v}\left[A(v,t)p(v,t \mid v_0,t_0)\right] \\ &\quad + \frac{1}{2} D \frac{\partial^2}{\partial v^2} p(v,t \mid v_0,t_0)\ , \end{aligned} \tag{3}$$

where $A(v,t)$ and D are drift and diffusion coefficients, respectively. One can proceed to solve the differential equation (3) directly, but alternatively, as a springboard for nontrivial applications, we obtain the conditional probability density $p(v, t \mid v_0, t_0)$ by expressing it as a path integral ,

$$p(v,t \mid v_0,t_0) = \int \exp\{-S[v(t)]\} \mathcal{D}[v(t)] \ . \tag{4}$$

The integration over $\mathcal{D}[v(t)]$ is over all possible trajectories or histories of the fluctuating relative membrane potential $v(t)$ from the initial value v_0 to the specified terminal value v. The effective action $S[v(t)]$ is given by (setting $t_0 = 0$),

$$S = \frac{1}{2D}\int_0^t [\dot{v} - A(v,\tau)]^2 d\tau \ . \tag{5}$$

EVALUATING THE PATH INTEGRAL

To evaluate Eq.(4), we make use of the general method first introduced by Hida and Streit for the Feynman path integral, also naturally applicable to the Feynman-Kac integral. First, we parametrize the histories in terms of the Brownian motion $B(t)$,

$$v(t) = v_0 + \sqrt{2D} B(t)\ , \tag{6}$$

and, with v_0 fixed, the time derivative of $v(t)$ is given by,

$$\begin{aligned}\dot{v} &= \sqrt{2D}\frac{dB(\tau)}{d\tau} \\ &= \sqrt{2D}\omega(\tau)\,.\end{aligned}$$

The Gaussian white noise variable $\omega(\tau)$ may be viewed as the velocity of the Brownian motion, $\omega(\tau) = dB/d\tau$. Alternatively, we can then write, $B(t) = \int_0^t \omega(\tau)\, d\tau$. The effective action (5) can now be written as,

$$\begin{aligned}S &= \int_0^t \omega^2(\tau)\, d\tau - \sqrt{\frac{2}{D}}\int_0^t \omega(\tau)A(v,\tau)\, d\tau \\ &\quad + \frac{1}{2D}\int_0^t A^2(v,\tau)d\tau\,. \end{aligned} \tag{7}$$

If we take the simple case of a time-dependent drift coefficient, i.e., $A(v,t) = \xi(t)$, we obtain the following exponential in (4) as a white noise functional,

$$\begin{aligned}\exp\{-S[v(t)]\} &= \exp\left[-\int_0^t \omega^2(\tau)\, d\tau + \sqrt{\frac{2}{D}}\int_0^t \omega(\tau)\xi(t)\, d\tau\right] \\ &\quad -\frac{1}{2D}\int_0^t \xi(t)^2\, d\tau\,. \end{aligned} \tag{8}$$

Note that the parametrization (6) does not fix the final endpoint at $v(t) = v$. We can do this using the Donsker delta function,

$$\delta(v(t) - v) = \delta\left(v_0 - v + \sqrt{2D}\int_0^t \omega(\tau)\, d\tau\right). \tag{9}$$

Furthermore, since the integrand in Eq.(4) is now a white noise functional, the integration over the paths $\mathcal{D}[v(t)]$ or $d^\infty v$ is transformed into an integration over the Gaussian white noise measure, $d\mu(\omega) = N_\omega \exp\left[-\frac{1}{2}\int \omega(s)^2\, ds\right] d^\infty\omega$, with N_ω a normalization factor. In particular, $d^\infty v \to N_\omega d^\infty \omega = \exp\left[\frac{1}{2}\int \omega(s)^2\, ds\right] d\mu(\omega)$. Using this measure, together with equations (8) and (9), for the path integral (4) we have,

$$\begin{aligned}p(v,t \mid v_0,t_0) &= \exp\left(-\frac{1}{2D}\int_0^t \xi^2(t)d\tau\right) \\ &\quad \times \int I_0 \exp\left(\sqrt{\frac{2}{D}}\int_0^t \omega(\tau)\xi(t)d\tau\right) \\ &\quad \times\delta\left(v_0 - v + \sqrt{2D}\int_0^t \omega(\tau)\, d\tau\right) \\ &\quad \times d\mu(\omega)\,, \end{aligned} \tag{10}$$

where I_0 is the Gauss kernel,

$$I_0 = N\exp\left(-\frac{1}{2}\int_0^t \omega^2(\tau)d\tau\right)\,, \tag{11}$$

with N an appropriate normalization. Eq.(10) can be simplified by expressing the Donsker delta function in terms of its Fourier representation to get,

$$p(v,t \mid v_0,t_0) = \exp\left(-\frac{1}{2D}\int_0^t \xi^2(t)d\tau\right) \times \frac{1}{2\pi}\int_{-\infty}^{+\infty} dk \exp\left[ik(v_0 - v)\right] \times \int I_0 \exp\left\{i\int_0^t \omega(\tau)\left[k\sqrt{2D} - i\sqrt{\frac{2}{D}}\xi(t)\right]d\tau\right\} d\mu(\omega). \tag{12}$$

Denoting, $\xi' = k\sqrt{2D} - i\sqrt{2/D}\xi(t)$, the integration over $d\mu(\omega)$ is identified as just the T-transform of the Gauss kernel I_0, i.e.,

$$TI_0(\xi') = \int \exp\left[i\int \omega(\tau)\xi'(\tau)\right] I_0 d\mu(\omega), \tag{13}$$

and we obtain the expression for the conditional probability,

$$p(y,t \mid y_0,t_0) = \frac{1}{2\pi}\int_{-\infty}^{+\infty} dk \exp\left(-k^2Dt/2\right) \times \exp\left\{ik\left[(v_0 - v) + \int \xi'(\tau)d\tau\right]\right\}. \tag{14}$$

The remaining integral is a Gaussian integral which can be evaluated to yield,

$$p(v,t \mid v_0,t_0) = \sqrt{\frac{1}{2\pi Dt}}\exp\left[-\frac{1}{2Dt}(v_0 - v)^2\right] \times \exp\left[-\frac{(v_0 - v)}{Dt}\int_0^t \boldsymbol{\xi}(\tau)d\tau\right] \times \exp\left\{-\frac{1}{2Dt}\left[\int_0^t \boldsymbol{\xi}(\tau)d\tau\right]^2\right\}. \tag{15}$$

DISCUSSION AND OUTLOOK

We have obtained an expression for the conditional probability density function for the relative membrane potential involved in neural activity for the case when there is a time-dependent drift. The drift introduces two factors that modulate the free probability density. With $v(t)$ being a voltage difference that triggers the firing of the neuron, the drift coefficient can be interpreted as the variation of the total current crossing the neuronal membrane. In an explicit solution for $p(v,t \mid v_0,t_0)$, the time-dependent drift coefficient $A(t) = \boldsymbol{\xi}(t)$ can be chosen to model certain features of brain processes. Periodic oscillations of cellular currents can be described by taking $A(t) = \sigma\cos(\omega t)$, a rapidly increasing change in current by $\boldsymbol{\xi}(t) = \lambda\exp(ct)$, or by the polynomial form, $\boldsymbol{\xi}(t) = \alpha t^s$, where σ, λ,and α are biochemical parameters. Such a polynomial dependence in time of the drift can already strongly damp the probability density function for the firing of a neuron. One could also look at voltage-dependent drift coefficients, $A(v,\tau)$. For this case, we would have interaction terms in the action of the path integral. This work, still in progress, would allow more room for modeling the complex interneural connections. This also leads to generalization to N neurons with path integrals in large-N-dimensions. Symbolically, such path integral may be written as,

$$P(\widehat{v},t \mid \widehat{v}_0,t_0) = \int e^{-S(\widehat{v})}\mathcal{D}[\widehat{v}]\ , \tag{16}$$

where, $\widehat{v}= (v_1, v_2, ..., v_N)$ and an effective action, $S(\widehat{v}) = \int \mathcal{L}\, dt$, with a Lagrangian,

$$\mathcal{L} = \frac{1}{2}\left(\frac{d\widehat{v}}{dt} - \mathbf{A}(\widehat{v}, t)\right)^T \mathbb{B}(\widehat{v}, t)^{-1}\left(\frac{d\widehat{v}}{dt} - \mathbf{A}(\widehat{v}, t)\right). \quad (17)$$

Here, $\mathbf{A} = (A_1, A_2, ..., A_N)$, where A_2 would be the drift coefficient (change in time of total current) for the neuron located at $\mathbf{x}_2$. Taking, $\mathbb{B} = D\mathbb{I}$, with $\mathbb{I}$ an identity matrix and D a constant diffusion coefficient, Eq. (16) becomes,

$$P(\widehat{v}, t+\tau \mid \widehat{v}_0, t) = \left(\frac{1}{2\pi D\tau}\right)^{3/2} \exp\left\{-\frac{1}{2D}\left[\left(\frac{\Delta\widehat{v}}{\tau}\right) - \mathbf{A}(\widehat{v}, t)\right]^2 \tau\right\}. \quad (18)$$

Another direction for study is to incorporate the observation that real brain tissue shows neurons embedded in a mass of glial cells such as astrocytes and oligodendrocytes. The proportion is roughly 1 neuron to 10 glial cells. Recent work has shown that such glial cells do not just play an ancillary role in the activity of neurons. Indeed, they are observed to enhance synaptic transmission of signals [12]. Moreover, it is also of interest to see how boundary conditions at cellular membranes can be handled within the framework of white noise path integrals. For example, mixed boundary conditions such as those discussed in [13] can be accommodated in the white noise path integral approach to obtain a closed form for the conditional probability density function.

REFERENCES

1. See e.g., S. Gilman and S. W. Newman, *Manter and Gatz's Essentials of Clinical Neuroanatomy and Neurophysiology*, 10th ed, F. A. Davis, 2002; B. Young and J. W. Heath, *Wheater's Functional Histology*, 3rd ed.
2. OECD-CERI, "Understanding the Brain: Towards a New Learning Science" (Organization for Economic Co-operation and Development, 2002).
3. Newly discovered is the human relaxin HR3. See R. A D. Bathgate et al., *Trends in Endocrinology and Metabolism* **14,** 207 (2003).
4. H. Koizumi et al, *Jour. Biomed. Opt.* **4,** 403 (1999).
5. See, e.g., J. J. Hopfield, in *More Things in Heaven and Earth: A Celebration of Physics at the Millennium*, edited by B. Bederson, Springer, APS, New York, 1999, p. 726.
6. See e.g., J. A. Kelso, *Dynamic Patterns: The Self-Organization of Brain and Behavior,* MIT Press, Cambridge, 1995.
7. K. Yasue, M. Jibu, T. Misawa, and J.-C. Zambrini, *Ann. Inst. Statist. Math.* **40**, 41-59 (1988).
8. R. E. Shaw, E. E. Kadar, and M. T. Turvey, *Behavioral and Brain Sciences* **20**:2, 265 (1997)
9. T. Ohira and J. D. Cowan, "Path Integrals for Stochastic Neurodynamics," TR-94-016, Sony, Tokyo, 1995.
10. C. C. Bernido and M. V. Carpio-Bernido, *Jour. Math. Phys.* **43,** 1728-1736 (2002) .
11. L. Streit and T. Hida, *Stoch. Proc. Appl.* **16**, 55-69 (1983) .
12. S. Nadkarni and P. Jung, *Phys. Biol.* **4**, 1-9 (2007).
13. R. Erban and S. J. Chapman, *Phys. Biol.* **4**, 16-28 (2007).

BIOLOGICAL STRUCTURES AND INTERACTIONS

Biological Structures, Interactions, Function and Behavior: Research Opportunities for Physicists

Gisela P. Concepcion

Marine Science Institute, University of the Philippines, Diliman, Quezon City 1101, Philippines

Abstract. Studies on marine biomolecules at the Marine Natural Products Laboratory (MNPL) and studies on biomedically relevant proteins at the Virtual Laboratory of Biomolecular Structures (VIRLS) of the University of the Philippines Marine Science Institute (UPMSI) are presented. These serve to illustrate some underlying principles of biological structures, interactions, function and behavior, and also to draw out some unresolved questions in biology of possible interest to non-biologists. The Biological Structures course offered at UPMSI, which aims to introduce underlying biological principles to non-biology majors and to promote trans-disciplinary research efforts, is also presented.

Keywords: biological structures, biological interactions, specificity, marine biodiversity, biomolecular structure analysis, vaccine design, biomedical applications

PACS: 87.10 – 87.23

INTRODUCTION

I am grateful for this opportunity to share with you some of our work and ideas, and perhaps propose ways for us to "cross borders and explore interfaces." As an overview, I will start by telling you about the Biological Structures course that we offer at the University of the Philippines Marine Science Institute (UPMSI). Then I will share with you what I think are some underlying principles of biological interactions; and then to illustrate these principles, I will talk briefly about our research projects on: 1) marine biodiversity as a source of potential drugs and related biomedical research, and 2) biomolecular structure analysis and vaccine design and development. In relation to these projects, I will pose some open or unresolved questions in biology that might be of interest to physicists, mathematicians, and computer scientists.

BIOLOGICAL STRUCTURES COURSE

Our Biological Structures course is patterned after the course that was offered by the microbiologist Salvador Luria at Massachusetts Institute of Technology in the middle of the last century. Our course was conceptualized by the structural biologist Eduardo Padlan, Adjunct Professor at UPMSI, and is team-taught by a group of faculty members at the University of the Philippines (UP) with different specializations. It covers a wide range of biological structures -- from small molecules, to macromolecules, macromolecular assemblies, cells, tissues, organisms, and ecosystems. To illustrate these

CP1021, *Stochastic and Quantum Dynamics of Biomolecular Systems, Proceedings of the 5th Jagna International Workshop,* edited by C. C. Bernido and M. V. Carpio-Bernido

structures, we take examples from the sea -- where life began. The recurring question is -- are there underlying principles, with fine similarities and differences, at these different levels of biological structures?

In Unit 1, I discuss the structure and function of small biomolecules -- secondary metabolites which are produced in minute quantities by lower organisms, e.g., sponges, the earliest metazoans, and their associated microorganisms, for defense and survival in the competitive marine environment. (The field of marine natural products or secondary metabolites is one of my research interests.) Primary metabolites are produced in larger quantities and play essential roles in running the cell's machinery. Lipids form inert membranes that are used to compartmentalize components in the cell. Steroids, with structures differing in rigidity and flexibility, can serve either as membrane structures or hormones, with the latter showing highly specific interactions. Glucose is a major source of energy but also of asymmetric or chiral carbon centers that can produce specific conformations capable of forming specific binding interactions. Carotenes are rigid, highly conjugated linear structures which are photo-reactive and participate in the process of color vision; while the highly conserved aromatic porphyrins are found in both chlorophyll and hemoglobin, and function to capture sunlight and bind oxygen, respectively [1].

For each of these small molecules, I discuss unique structural features and physico-chemical properties that make them suitable for a basic cellular function and thus have been conserved through evolution. The rest of the biological structures discussed in the course are presented in Table 1. For each of the structural levels, certain underlying concepts pertaining to structure, statics, dynamics, function and behavior are discussed where they are pertinent or applicable (Table 2). There have been many UP graduate students in physics, mathematics and computer science interested in biology who have enrolled in this course.

TABLE 1. Biological structures course outline.

Unit 1: Small Molecules Primary and Secondary Metabolites Lipids Steroids Carbohydrates Carotenes Porphyrins	Unit 5: Marine Plants Microalgae Macroalgae Other Plants
Unit 2: Macromolecules and Assemblies Nucleic Acids Proteins: Hemoglobins, Prions, Antibodies Macromolecular Complexes Viroids Viruses	Unit 6: Marine Animals Invertebrates Fish Other Vertebrates
Unit 3: Cells: Unicellular and Simple Organisms Mycoplasms Bacteria Protozoa Other Lower Eukaryotes	Unit 7: Marine Communities and Ecosystems Benthic Communities Pelagic Communities Coral Reef Ecosystems Mangrove Ecosystems

Unit 4: Organ Systems in Higher Organisms Invertebrates Vertebrates Mammalian Systems Human Systems Plants	

TABLE 2. Some underlying concepts of biological structures.

STRUCTURE

Statics

- Structural elements
  - Units and interactions between units
  - Stability and energetics of interactions
- Internal structure
  - Skeleton or framework
  - Relation to external structure
- External structure
  - Form or appearance, surface features
  - Relation to internal structure
- Overall design and strategy
  - Structural organization and integrity
  - Purpose or function of structure

Statics/Dynamics

- Bioenergetics and stability
  - Covalent and noncovalent interactions
  - Stereochemistry, conformation, folding
  - Conformational change, allosteric control, structural modification
  - Macromolecular interactions and complexes
- Compartmentalization, tethering, localization, translocalization
- Synthesis, degradation, death
- Cycling, turnover, recycling
- Internalization, endocytosis, exocytosis, secretion, excretion

FUNCTION AND BEHAVIOR

Internal Dynamics

- Specificity, recognition, adaptability
- Feedback and feedforward regulation and control, switching and sigmoidicity
- Pleiotrophism, homeostasis
- Parsimony versus waste
- Gene duplication, amplification, recombination, alternative splicing, epigenetics
- Protein-protein, protein-gene, protein-ligand interactions and signaling networks

Dynamics in Response to the Environment

- Survival, defense, competition, predation, deterrence
- Metabolism, growth, reproduction
- Cycling, apoptosis, degradation and death

Evolutionary Dynamics

- Evolutionary rate/mutation rate
- Adaptation, escape mechanisms
- Pattern of development
- Convergent or divergent evolution
- Direction and purpose of evolutionary change

SOME UNDERLYING PRINCIPLES OF BIOLOGICAL STRUCTURES AND INTERACTIONS

An important underlying principle of biological interactions, according to Dr. Padlan, seems to be that "...atoms, molecules, cells, and even individual organisms act in accordance with the basic notion of attraction versus repulsion, group bonding versus individuality, cooperativity and synergy versus independent action, enthalpy versus entropy..."

Specificity is the hallmark of stable biological interactions. And I think recognition and regulation are the hallmarks of biological and evolutionary change. Recognition occurs at the present time, but the biological system has memory -- in the form of regulatory modes or controls -- already in place. Upon molecular recognition brought about by specific physicochemical interactions, there could be drastic change, but the interaction eventually undergoes regulation or modulation, and homeostasis/balance and stability are regained. Sigmoidicity is a typical dose-response and time-response curve and indicates how a biological system changes -- adjusts or switches -- from one state to another. Programmed regulation results in slow, evolutionary change and safeguards against the prolonged, destabilizing or catastrophic effects of drastic changes.

But exactly how does molecular recognition occur between macromolecules, considering that in the cell, macromolecules are found at high concentrations? How do proteins seek out and recognize their partners in the crowded cytoplasmic 'soup'? How do proteins interact with one another specifically and sequentially to transmit a signal from the cell's exterior to the cytoplasm to the nucleus? These are some questions that are being addressed presently and are not yet fully answered.

OUR RESEARCH PROJECTS: AN OVERVIEW

I would now like to introduce our research projects where I think there might be research opportunities for physicists, mathematicians, and computer scientists:

1. In collaboration with Eduardo Padlan, protein crystallographer, structural biologist and molecular immunologist:

- Antibody-targeted therapy:
  - anticancer immuno-liposomal-drug: biophysicochemical characterization, *in vitro* and *in vivo* testing
- Biomolecular structure and sequence analysis:
  - prions: misfolding, infectivity, transmissibility, natural function
  - autoimmune antibodies: structural correlates of affinity maturation
- Vaccine design and development:
  - human and avian flu
  - malaria
  - dengue

2. In collaboration with Eduardo Mendoza, mathematician and computer scientist:
- Petri Net modeling of apoptosis pathways and pro-apoptotic mechanisms of cytotoxic marine compounds

3. In collaboration with Baldomero Olivera, biochemist and marine neuroscientist:
- Molecular phylogeny, rapidly diversifying toxin genes and neuroactive peptides of mollusks

Our research projects are focused on an area of competitive advantage that we have in our country -- marine and terrestrial biodiversity. This biodiversity includes bioactive compounds, biomaterials, biocatalysts and also disease molecular targets from disease tissues such as cancer and infectious diseases. Our molecular studies involve small, medium and large biomolecules.

The priority medical conditions we are interested in are: cancer, infectious diseases (Gram positive and Gram negative bacterial infections, tuberculosis, malaria, influenza and dengue), inflammation and cardiovascular disease (platelet aggregation), transplantation (immunosuppression), and pain (cancer pain and neuropathic pain).

MARINE BIODIVERSITY AS A SOURCE OF POTENTIAL DRUGS AND RELATED BIOMEDICAL RESEARCH

Bioactive Marine Natural Products

Over the last fifteen years, the Marine Natural Products Laboratory or MNPL at the UPMSI, in collaboration with Chris Ireland of the University of Utah, has been involved in the isolation of bioactive marine compounds from marine invertebrate organisms, e.g., sponges and ascidians, and marine microorganisms (Table 3). The chemical diversity of the compounds (Fig. 1) reflects the biodiversity of the source organisms which in turn reflects the adaptations of the organisms to the environment. The organisms produce an array of compounds depending on their specific growth, defense and survival requirements.

Marine biodiversity and chemodiversity indeed represent a largely untapped pharmacological resource. But while a number of these compounds have been shown to possess significant anticancer or anti-infective activity, toxicity to normal mammalian cells *in vitro* and *in vivo* remains the major hindrance to their development as useful drugs. Marine cytotoxins owe their toxicity to their nonspecificity which is due to the presence of multiple pharmacophores in their structure resulting in binding to different parts of a cell. *In vivo*, a pharmacophore may interact with multiple, unintended molecular targets found in a variety of cells (nonselectivity), also resulting in toxicity. In the case of cancer, normal and cancer cells may share the same or similar molecular targets and so marine cytotoxins may also kill normal cells. In some cases, a cancer cell simply overexpresses a molecular target found in a normal cell.

TABLE 3. Some bioactive marine compounds isolated from Philippine marine organisms

Compounds	**Bioactivity**	**Reference**
Crambescidin 800 and synthetic analogues	Antimalarial *in vivo*	Lazaro et al. *J Antibiot*, 2006 [2]
Kalihinols	Bacterial folate biosynthesis inhibitors	Bugni et al. *Tetrahedron*, 2004 [3]
Microcionamides A and B	Cytotoxic against human breast cancer cell lines	Davis et al. *J Org Chem*, 2004 [4]
Cribrostatin, renierone, O-demethylrenierone	Cytotoxic against a human colon cancer cell line	Sandoval et al. *Nat Prod Res*, 2004 [5]
Lissoclinotoxins E and F	Cytotoxic against a human breast cancer cell line	Davis et al. *Tetrahedron*, 2003 [6]
Xestospongin/araguspongine-type molecule and 5,6-dibromotryptamine	Platelet aggregation inhibitors *in vitro*	Pimentel et al. *Mar Biotechnol*, 2003 [7]
Neoamphimedine	Cytotoxic (topoisomerase II-mediated) and anti-neoplastic *in vivo*	Marshall et al. *Biochem Pharmacol*, 2003 [8]; de Guzman et al. *J Org Chem*, 1999 [9]
Heptyl prodigiosin	Antimalarial *in vivo*	Lazaro et al. *J Nat Toxins*, 2002 [10]
Bromoindole derivatives and terpenes	Cytotoxic against a human colon cancer cell line	Tasdemir et al. *Z Naturforsch [C]*, 2002 [11]
Perophoramidine	Cytotoxic against a human colon cancer cell line; induces apoptosis	Verbitski et al. *J Org Chem*, 2002 [12]
10Z-hymenaldisine and hymenin	MEK-1 inhibitors	Tasdemir et al. *J Med Chem*, 2002 [13]
Stelletins B and E	Cytotoxic against a human colon cancer cell line	Tasdemir et al. *J Nat Prod*, 2002 [14]
Agelasine F	Anti-TB *in vitro*	Mangalindan et al. *Planta Med*, 2000 [15]
Makaluvamine N	Topoisomerase II-inhibitor	Venables et al. *J Nat Prod*, 1997 [16]
Adociaquinone B	Topoisomerase II-inhibitor	Concepcion et al. *J Med Chem*, 1995 [17]

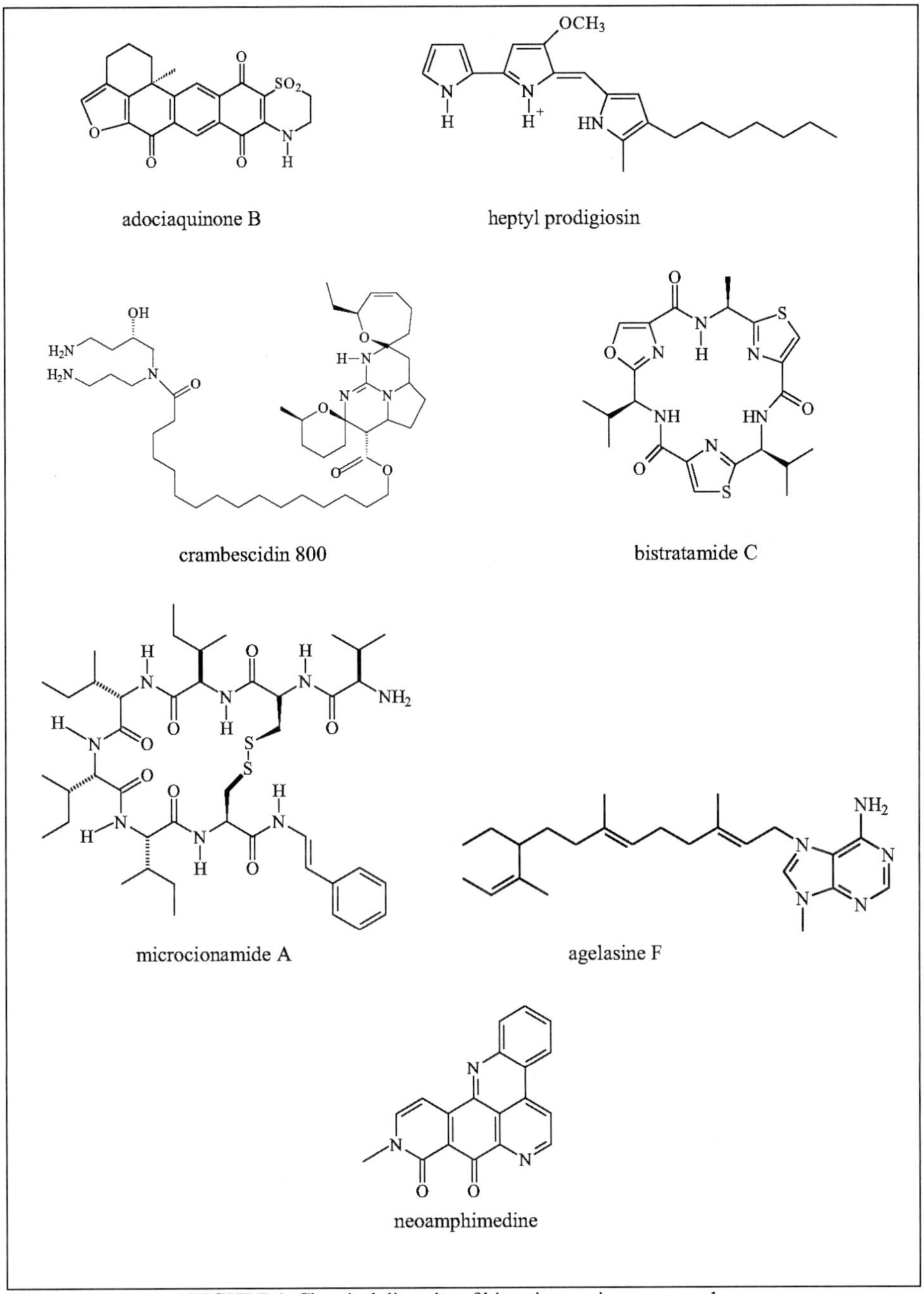

FIGURE 1. Chemical diversity of bioactive marine compounds

Development of Targeted Drug Therapy

The problems of nonspecificity and toxicity can be addressed in a number of ways, but principally by devising a way to target the marine cytotoxin to the disease cell -- like a molecular-guided missile. In 2000, in collaboration with Eduardo Padlan, who was involved in the development of humanized antitumor antibodies at the US National Institutes of Health, we pursued the development of antibody-targeted drug therapy for breast cancer under the government-funded project called AMOR (Antibody and Molecular Oncology Research). Marine compounds with low molecular weights (300-450 m/z), high cytotoxicity (nanomolar IC_{50}), steep dose-response curves, and identified molecular targets, were loaded onto a drug delivery system which was chemically linked to an antitumor antibody fragment. The antibody used was one that was generated against a pancarcinomic tumor marker found on the surface of cancer cells and thus, the antibody recognizes the tumor marker specifically (Fig. 2) [18].

In preliminary studies, tumor regression in an animal tumor model was demonstrated as proof of principle or proof of efficacy. The AMOR product prototype represents an alternative to the classical regimen for treating cancer which often involves any single treatment or any combination of the following methods: surgery ("slash"), radiation ("burn"), and chemotherapy ("poison"). Because the target tumor marker is not found on the surface of normal cells, the AMOR product spares normal cells and is nontoxic. Because it makes use of a 'humanized' antitumor antibody (mouse antibody is engineered to become or to appear similar to a human antibody), it does not cause allergy and is thus suitable for protracted therapy.

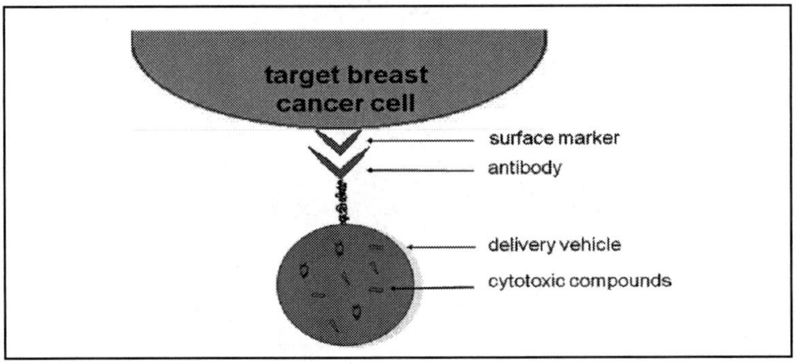

FIGURE 2. Design of targeted anticancer therapy

As a sequel to AMOR, biophysicists are working closely with molecular biologists on the bio-imaging of tumor progression and of targeted therapy in an animal model using fluorescent proteins and nanodots [19]. And experimentalists and mathematical modelers together are trying to understand cancer circuitry, the evasion of apoptosis (a kind of programmed cell death) in cancer, and the effects of pro-apoptotic drugs on cancer cells [20,21].

Molecular Targets in Cancer and Modeling of Cellular Pathways

Cancer consists of cells which have acquired new capabilities compared to normal cells, such as self-sufficiency in growth signaling, insensitivity to anti-growth signaling, evasion of apoptosis, sustained angiogenesis, tissue invasion and metastasis, and limitless replicative potential leading to immortalization [22]. The underlying molecular mechanisms of these acquired capabilities are being gradually elucidated. The cell possesses an intricate circuitry involving protein-protein, protein-gene, small molecule-protein, and small molecule-gene regulatory networks. Cross-talking and feedback and feedforward loops enable the cell to react sensitively, quickly and synchronously to changes in its interior and exterior. In different types of cancer, different key proteins in the cell circuitry, acting as sensors, transducers, adaptors, rheostats, or amplifiers, are mutated. They behave abnormally and disrupt the normal processes in the cell, resulting in uncontrolled growth and proliferation and aggressive invasion [23].

In one study, the compound heptyl prodigiosin isolated from a marine microorganism in our laboratory was shown to induce apoptosis by targeting an apoptotic pathway upstream [20]. Experimental data obtained in this project will be used to create a mathematical model of the pro-apoptotic effect of the compound in order to provide more insights on apoptosis and to assist in the design of new experiments that can be used to further assess its anticancer potential.

The modeling of cancer pathways makes use of tools and strategies used in systems biology. An ongoing project involves the qualitative Petri Net (PN) modeling of the pro-apoptotic anticancer effects of the marine compound cephalostatin 1 [21,24]. Enzymatic, elementary or non-general reactions in the apoptotic pathway are represented using PN symbols and nomenclature which indicate the initial quantities of materials (reactants and products) called 'places' undergoing stoichiometric 'transitions' (reactions, interactions, processes) [25]. The study attempts further to extend the PN model to a quantitative, kinetic model using certain experimental or estimated kinetic parameters.

Neuroactive Peptides from Venomous Marine Snails

The work of Baldomero Olivera and his group in the last thirty years on neuroactive peptides from the venomous *Conus* snails is widely documented [26-29]. There are presently several conopeptides in preclinical and clinical trials for pain, myocardial infarction, and epilepsy. The first conopeptide to reach the market is Prialt (Primary Alternative to Morphine), also known as Ziconitide, which was approved by the US FDA in 2004 for the treatment of intractable pain in cancer, and for severe, chronic neuropathic pain. Prialt is a chemically synthesized version of the native peptide ω-conotoxin MVIIA (SNX-111) that Dr. Olivera's group isolated from the venom of *Conus magus* from the Philippines in the mid-80s [30]. It was initially described as a 'shaker' peptide, and later found to be a selective blocker of N-type neuronal voltage-gated calcium channels [31].

Conopeptides such as MVIIA are small peptides (~15-30 amino acids long), many of which are disulfide-rich. The disulfide bridges serve to secure a tightly and specifically folded structure of the conopeptides, which are able to bind specific sites on the surface of neuronal or muscle membrane receptors known as ion channels, e.g., Na^+, Ca^{2+}, K^+ channels, acetylcholine and other nicotinic acid receptors, N-methyl-D-aspartate (NMDA) receptors. The binding causes the blocking of the movement of ions through the channel which then impairs normal nerve and/or muscle function, e.g., abrogation of the action potential leading to a variety of effects such as paralysis, excitotoxic shock, sedation, etc.

Venomous snails have evolved to become the most megadiverse and among the most successful organisms in the marine environment. The cone snails capture their prey, i.e., fish, worms, or other snails, using venom delivered through a hollow arrow or harpoon. The combination of conopeptides produced in the venom is determined by the particular feeding strategy the snail uses to catch and paralyze its prey, and to deter predators and competitors. This strategy represents one of the most efficient examples in Nature of how organisms can quickly adapt to their environment by gene expression. The snails rapidly diversify or mutate the 'exogenes' coding for venom peptides (which are direct gene products, although some are posttranslationally modified) as their adaptive response to their prey and the conditions in the environment.

Conopeptide genes are translated as precursor proteins consisting of three regions: a highly conserved *pre* region which codes for a hydrophobic signal sequence that leads the peptide to the surface of the cell for secretion, a semi-conserved *pro* region whose function is still unclear, and a mature sequence that contains a conserved cysteine scaffold surrounded by hypermutable sequences. The *prepro* sequence is eventually cleaved at a specific site yielding the mature sequence which is where the neuroactive property resides.

Exactly how the mature conopeptide folds in a specific manner, however, with fixed cysteine partners forming the disulfide bonds, is still very little understood. Generally, it takes ~50 amino acids in a protein to fold into a domain or supersecondary structure which is the unit structure usually associated with a specific function, e.g., a catalytic domain. Furthermore, if there are six cysteine residues, they can form three disulfide bonds in 15 different ways. Disulfide isomerase, which is found present in the venom duct, catalyzes the reversible breaking and forming (rearrangement) of disulfide bonds, but it does not act to shift the equilibrium towards a particular disulfide linkage pattern. It has been postulated that a chaperone protein, interacting with the *pro* region of the propeptide, aids in its proper folding prior to the removal of the *pro* sequence. The conopeptide folding problem remains unresolved and would benefit from more biochemical, bioenergetics and kinetic studies.

The phylogenetic tree of the superfamily of Conoidean snails is being constructed based on the analysis of 16S rDNA and other molecular phylogenetic markers, and this is being used to verify earlier taxonomic identifications of closely related species based on shell morphology (Fig. 3). We have also recently initiated work on the molecular phylogeny of turrids (family Turridae), another group of snails closely related to *Conus*. While there are an estimated 500-700 *Conus* species, there exists probably >10,000 species of turrids (accounting for >90% of Conoidean biodiversity, with the center of turrid biodiversity identified to be in the Philippines) with a large fraction of species that remain undescribed and unexplored.

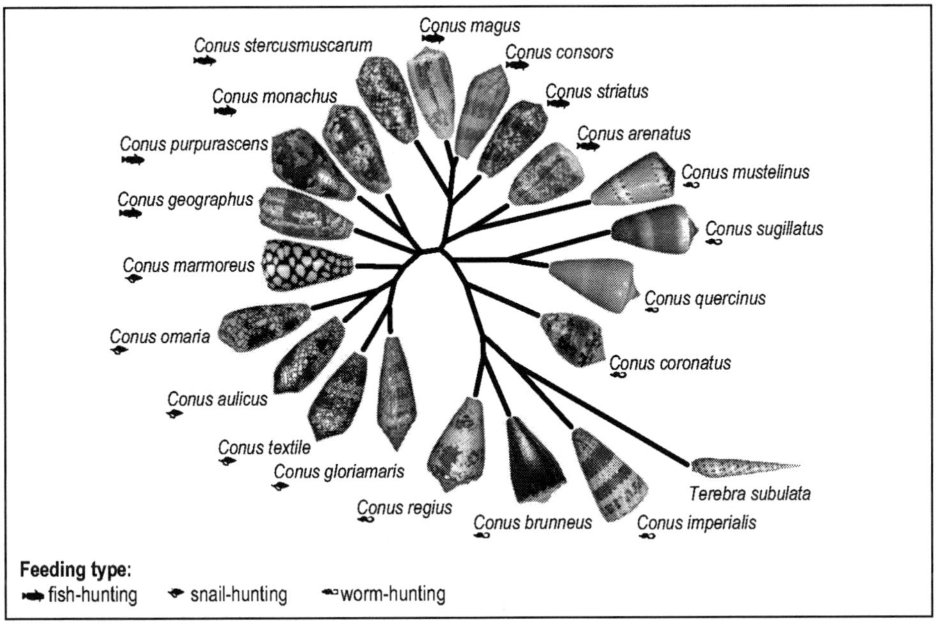

FIGURE 3. Phylogenetic position of *Conus* species based on 16s rDNA sequences

Following the *Conus* model, we are also attempting to identify novel toxin genes and purify toxin peptides from selected turrids [32, 33]. Turrids feed on polychaete worms and use their toxins to sedate, paralyze and feed on their prey. However, the major challenge in turrid research is obtaining a sufficient quantity of genetic material and venom, because most turrids are much smaller than *Conus* snails. Turrids requires excellent dissection tools and skills, and substantial manpower to dissect hundreds of them. One approach being taken is to isolate and sequence turritoxin genes, to clone them in a suitable expression system, e.g., bacteria, yeast, insect cells, and to produce recombinant peptide toxins in culture. In this way, we would circumvent the problem of having to acquire a lot of material and there would be no need for massive recollection of turrid samples. Ultimately, we will be able to conserve the supply of turrids in the wild.

Biomolecular Structure Analysis

The Virtual Laboratory of Biomolecular Structures (VIRLS) based in UPMSI and led by Dr. Padlan and the author is involved in the analysis of the sequences and structures of several proteins that are relevant to health and disease. Changes in the primary, secondary and tertiary structures of proteins would account for changes in their interactions, behavior and function.

Prions are ubiquitously occurring proteins whose normal function is still little understood. Prions become pathogenic when they misfold or change their secondary structure and acquire more β-sheet (and less α–helix) character. Infectious prions and familial pathogenic forms of prions lead invariably to fatal prion diseases such as 'mad cow disease' or bovine spongiform encephalopathy (BSE) in cattle, scrapie in sheep and goats, and kuru and Creutzfeldt-Jakob Disease (CJD) in humans.

In a series of papers, we performed sequence analysis of prions to explain misfolding, infectivity, transmissibility and genetic susceptibility. We compared the sequences of human and various animal prion proteins with regards to the fragments that could result from gastric digestion, and noted the close similarity of human and rodent prion sequences. We then proposed that a possible mode of BSE transmission to humans is not through cows but rodents [34]. In another study, we compared prion sequence variants in sheep differing in their susceptibilities to scrapie, by analyzing their secondary structures (β-sheet and α-helix content), and then compared these with human prion sequences. We proposed that direct transmission of scrapie from sheep to humans has not been observed, despite mutton being consumed by humans for centuries in many parts of the world, because the human prion protein is most similar to the sheep scrapie-resistant variant [35].

In another study, we hypothesized that in some familial cases, the V129M mutation in prions accounts for a higher incidence of prion disease in individuals homozygous for M129 because the codon for methionine at this position provides an alternative initiation site for translation resulting in an N-terminal truncated form that is missing the signal peptide. Thus the prion molecule is not transported to the extraceullular space but rather accumulates in the cytosol where it is more susceptible to fibril formation and aggregation [36].

The interactions, function and behavior of a protein are studied based on an accurate model of the 3D structure of the protein. In secondary, supersecondary and tertiary structure prediction studies, the best algorithms predict structure at less than 90% accuracy compared with high-resolution x-ray crystal structures [37, 38]. The development of neural network-based 3D structure prediction algorithms continues to be an important and challenging field. In these algorithms, the propensity of an amino acid to form or break an α-helix, β-sheet or a random coil is quantified based on its occurrence in secondary structures that are found in a reference group of proteins for which there are high-resolution x-ray crystal structures available. Because of the numerous biomedical applications of protein design, protein engineering and protein targeting, there is a great need to develop better algorithms that can predict protein structures very accurately e.g., proteins can be grouped into different structural classes, so that the propensity values can be determined more accurately for each class.

The structural correlates of affinity maturation of antibodies were studied by analyzing the sequences of 80 high-affinity autoimmune IgG antibodies to TPO (thyroxine peroxidase) [39]. Mutations in the 80 high affinity sequences derived from six germlines were analyzed in terms of the physicochemical properties of the replacement residues, i.e., hydrophilicity, size and polarizability, and charge and polarity, in the context of residue position and probable solvent accessibility. The effects of these substitutions were evaluated in terms of the resultant increased chemical interactivity potential of the affinity-matured antibodies relative to the germline (for each amino acid position along the framework regions (FRs) and complementarity-determining regions (CDRs) of the variable light (V_L) and variable heavy (V_H) chains). Antibody affinity was shown to be a function of chemical interactions, structural plasticity and stability.

Noncovalent interactions such as hydrogen bonding, *Van der Waals* interactions, polar interactions, electrostatic interactions, and hydrophobic effects, are responsible for the specific interactions between the antigen binding fragment of an antibody and the antigen, as well as for other protein-protein and protein-DNA surface interactions. Noncovalent interactions have much lower binding energies (1-10 kcal/mole) than covalent bonds (100 to 200 kcal/mole), but these interactions, when combined to act within a substantial surface area, are significant in strength and thus are effective. In the case of the IgG type of antibody, specific interactions of the CDRs of the antibody with an epitope of the antigen causes conformational changes in the antibody structure which result in the recruitment by the antibody of phagocytic macrophages to the site surrounding the antigen [40, 41].

Specific noncovalent interactions are also critical in maintaining the internal structure of a protein. While it is fairly straightforward to measure interaction energies on the surface of proteins, it is much more difficult to accurately measure interactions in the interior. One consideration is that the dielectric constant of water, which is given the standard value of 79 on the exterior, is grossly estimated to be ~4 in the interior where water molecules may be trapped. The dielectric constant of water would affect significantly the strength of interactions both in the exterior and interior of the protein. However, the dielectric constant value of water in the interior remains undetermined experimentally.

An interesting point to reiterate is that a cell consists of a very concentrated aqueous solution of proteins and other macromolecules. Determinants of specific protein folding motifs and specific protein interactions would be bioenergetics, including Brownian movement or thermal motion originating from heat generated by metabolic reactions and dissipated to the surrounding water molecules. [Metabolically coupled biochemical reactions produce or use chemical energy, e.g., in the form of ATP, but some chemical energy is inevitably lost (not caught completely by ATP or other molecules in the transfer) and irreversibly converted to heat energy or non-useful, entropic energy. The temperature of the human body or an organism, maintained by water, which has a high specific heat, reflects the heat or entropic content.] However, I think Brownian movement cannot fully account for the dynamics of protein folding and protein interactions. The challenge for a protein is to be able to assume a finely-tuned, properly folded 3D structure and then to be able to function properly, e.g., by finding its specific interacting partner(s), e.g., as part of a signaling pathway.

Vaccine Design and Development

Finally, I would like to tell you about the other major research interest of our group. This is to develop recombinant protein vaccines that would protect us from constantly mutating pathogens such as the influenza virus, cold virus, AIDS virus and the malaria parasite. The work is based on the rational design and protein engineering of vaccines by Dr. Padlan which takes into account the highly mutable nature of the pathogens [42]. We aim to develop vaccines that would work against different strains of the pathogens that emerge during different years or seasons. We have recently received government funding to develop a human influenza A vaccine. We will be asked to develop an avian influenza vaccine if the need arises, although presently the Philippines is avian influenza-free.

The problem with this season's influenza vaccine is that it may not work next season. Some pathogens evade the immune system by constantly mutating. In so doing, they manage to render the antibodies, which had been generated against previous strains, no longer totally protective. In many instances, the pathogens are able to make the evasive move by mutating a few amino acids, sometimes as few as one, in a surface molecule that may contain several hundred residues.

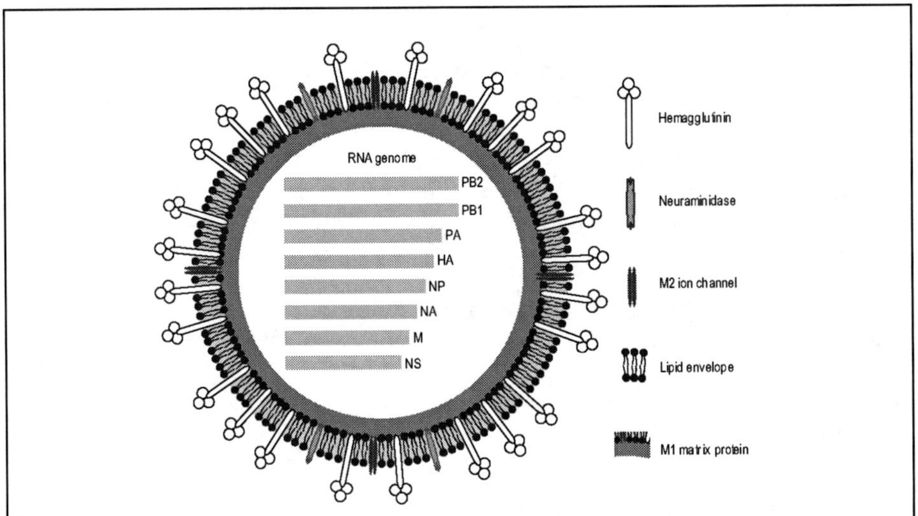

FIGURE 4. Schematic representation of influenza A virus. The genome of influenza A virus consists of 8 segmented single-stranded negative sense RNAs that code for 11 proteins. Three of which code for polymerases (PB2, PB1, and PA); while two code for surface glycoproteins H or HA (hemagglutinin) and N or NA (neuraminidase). The other proteins produced are NP (nucleoprotein), M1 (matrix protein), NS1 (non-structural protein), NEP (nuclear export protein), and the recently discovered PB1-F2 which is encoded by a frame-shifted reading of PB1 gene.

In the influenza virus, hemagglutinin (H) and neuraminidase (N) are the targets of antibodies in the immune response to influenza virus. In influenza A, 16 H and 9 N subtypes have been identified (Fig. 4) [43]. A particular strain is identified by its H and N subtypes, e.g., the current subtype of avian influenza is H5N1; the current circulating subtype of human influenza A in the Philippines is H3N2. If we could figure out precisely how the influenza virus manages to evade the immune system by making just a few amino acid changes, we may be able to design a vaccine that would be effective even if the virus is constantly mutating.

Antibodies can be generated against any accessible part of any macromolecule. But, while any accessible part of a macromolecule could elicit an antibody response, some parts are more antigenic than others. Those are the 'immunodominant epitopes' (IDEs) and most of the antibodies elicited by the antigen will be directed against those epitopes.

Influenza virus manages to evade the immune system by making amino acid changes in its IDEs. And so a possible strategy against constantly mutating pathogens is the 'de-Antigenization' of the IDEs. The rationale for this is that a modifed antigen, in which the IDEs had been rendered less antigenic, will elicit an antibody response to other parts of the antigen. A de-Antigenized antigen would then be an effective vaccine. The pathogen could continue to mutate its IDEs, but the vaccinated individual will have antibodies able to bind to the pathogen's other parts.

Our first choice of candidate vaccine is based on H because it is found to be present five times more than N on the surface of the virus. H has critical functions in the virulence pathway, being responsible for initial binding to the sialic acid residues on the surface of the host cell. In the endosome, at lower pH, H changes conformation and allows the opening up of the viral lipoprotein coat/capsid leading to the entry of ribonucleoprotein into the host cell.

A vaccine based on de-Antigenized N has also been designed. N is an enzyme that cleaves the sialic acid residues on the surface of the host cell to free up newly assembled viral particles and aid in their release from the cell and the infection of other cells. A vaccine combination of de-Antigenized H and N would probably be best but an H vaccine would be sufficient.

To 'de-Antigenize', one must get a good 3D crystal structure or 3D model of the protein. Then an antigenicity plot is generated along the amino acid sequence based on the physicochemical properties of the amino acids and the 3D structure. This allows us to determine the IDEs which are compared with IDEs identified from experimental data. Then the antigenicity of the IDEs is reduced by replacing some amino acids judiciously, while preserving the 3D structure of the protein, and a de-Antigenized plot is generated.

A vaccine for influenza is designed to be suitable to a particular country based on an analysis of the predominant strains that have been found in that country. Vaccines have been designed for Philippine influenza strains, Hongkong strains, Bangkok strains and Indonesian strains. Dr. Padlan has applied for a patent for the de-Antigenization technology in the USA. The technology is being implemented in the Philippines through projects at the UPMSI funded by the Philippine government and through a private company Vivotech Labs, Inc. which is commercializing Dr. Padlan's technology and accepts protein design and engineering contracts.

While the success of a scientist has long been judged by the number and quality of publications, presently, commercializable technologies protected by patents and other forms of intellectual property are fast becoming the tangible results that good basic science is expected to generate, especially if they can be used to address serious societal problems, such as public health. Indeed, the challenge to beat the mutating strains of influenza virus by developing a 'universal' vaccine is worthy of a creative and sustained scientific endeavor which, if successful, deserves not only scientific recognition but also financial renumeration.

CONCLUSIONS AND DIRECTIONS

In all aspects of our work on the study of the structure, function, interactions and behavior of small, medium and large biomolecules, there is a great need for trans-disciplinary efforts involving not only ecologists, biologists, biochemists, molecular biologists and pharmacologists, but also chemists, physicists, mathematicians, computer scientists and engineers. Biology is expectedly the most complex and interesting subject of scrutiny by the physical and mathematical sciences. Theoretical physicists, mathematicians and computer scientists should pursue theoretical calculations, predictive modeling and rational design based on empirical and experimental evidence, and in an iterative process, continue refining their models by collaborating with experimentalists who can experimentally validate the models. Non-biologists are now considered also major players in biology and are strongly encouraged to contribute significantly to the further understanding of underlying principles of biological systems – of biological structure, interactions, function and behavior.

ACKNOWLEDGEMENTS

I am grateful to my research associates in the Marine Natural Products Laboratory, namely: John Phillip Fadul, for his assistance with the text and references; Jan-Michael Cayme, for the chemical structures in Fig. 1; and Marvin Altamia, for the schematic representations in Figs. 3 and 4.

REFERENCES

1. J.H. Fuhrhop and C. Endisch. *Molecular and Supramolecular Chemistry of Natural Products and Their Model Compounds*. Marcel Dekker, Inc., New York and Basel, 2000, 602 pp.

2. J.E.H. Lazaro, J. Nitcheu, N. Mahmoudi, J.A. Ibana, G.C. Mangalindan, G.P. Black, A.G. Howard-Jones, C.G. Moore, D.A. Thomas, D. Mazier, C.M. Ireland, G.P. Concepcion, P.J. Murphy and B. Diquet, *J. Antibiot.* **59,** 583-590 (2006).
3. T.S. Bugni, M.P. Singh, L. Chen, D.A. Arias, M.K. Harper, M. Greenstein, W.M. Maiese, G.P. Concepcion, G.C. Mangalindan and C.M. Ireland. *Tetrahedron* **60,** 6981-6988 (2004).
4. R.A. Davis, G.C. Mangalindan, Z.P. Bojo, R.R. Antemano, N.O. Rodriguez, G.P. Concepcion, S.C. Samson, D.D. de Guzman, L.J. Cruz, D. Tasdemir, M.K. Harper, X. Feng, G.T. Carter and C.M. Ireland, *J. Org. Chem.* **69,** 4170-4176 (2004).
5. I.T. Sandoval, R.A. Davis, T.S. Bugni, G.P. Concepcion, M.K. Harper and C.M. Ireland, *Nat. Prod. Res.* **18,** 89-93 (2004).
6. R.A. Davis, I.T. Sandoval, G.P. Concepcion, R.M. da Rocha and C.M. Ireland, *Tetrahedron* **59,** 2855-2859 (2003).
7. S.M.V. Pimentel, Z.P. Bojo, A.V.D. Roberto, J.E.H. Lazaro, G.C. Mangalindan, L.M. Florentino, P. Lim-Navarro, D. Tasdemir, C.M. Ireland and G.P. Concepcion, *Mar. Biotechnol.* **5,** 395-400 (2003).
8. K.M. Marshall, S.S. Matsumoto, J.A. Holden, G.P. Concepcion, D. Tasdemir, C.M. Ireland and L.R. Barrows, *Biochem. Pharmacol.* **66,** 447-458 (2003).
9. F.S. de Guzman, B. Carte, N. Troupe, D.J. Faulkner, M.K. Harper, G.P. Concepcion, G.C. Mangalindan, S.S. Matsumoto, L.R. Barrows and C.M. Ireland, *J. Org. Chem.* **64,** 1400-1402 (1999).
10. J.E.H. Lazaro, R.Z. Predicala, G.C. Mangalindan, J. Nitcheu, F. Nesslany, D. Marzin, G.P. Concepcion and B. Diquet, *J. Nat. Toxins* **11,** 367-368 (2002).
11. D. Tasdemir, T.S. Bungi, G.C. Mangalindan, G.P. Concepcion, M.K. Harper and C.M. Ireland, *Z. Naturforsch. [C]* **57,** 914-922 (2002).
12. S.M. Verbitski, C.L. Mayne, R.A. Davis, G.P. Concepcion and C.M. Ireland, *J. Org. Chem.* **67,** 7124-7126 (2002).
13. D. Tasdemir, R. Mallon, M. Greenstein, L.R. Feldberg, S.C. Kim, K. Collins, D. Wojciechowicz, G.C. Mangalindan, G.P. Concepcion, M.K. Harper and C.M. Ireland, *J. Med. Chem.* **45,** 529-532 (2002).
14. D. Tasdemir, G.C. Mangalindan, G.P. Concepcion, S.M. Verbitski, S. Rabidran, M. Miranda, M. Greenstein, J.N.A. Hooper, M.K. Harper and C.M. Ireland, *J. Nat. Prod.* **65,** 210-214 (2002).
15. G.C. Mangalindan, M. Talaue, L.C. Cruz, S. Franzblau, A. Richardson, C.M. Ireland and G.P. Concepcion, *Planta Med.* **66,** 364-365 (2000).
16. D.A. Venables, G.P. Concepcion, S.S. Matsumoto, L.R. Barrows and C.M. Ireland, *J. Nat. Prod.* **60,** 408-410 (1997).
17. G.P. Concepcion, T.A. Foderaro, G.S. Eldredge, J. Clardy, E. Lobkovsky, L.R. Barrows and C.M. Ireland, *J. Med. Chem.* **38,** 4503-4507 (1995).
18. G.P. Concepcion and A.D. Santos. *Antibody Biotechnology and Liposome Drug Delivery Technology for Experimental Therapeutics of Breast Cancer, Terminal Report,* Department of Science and Technology – Philippine Council for Health Research and Development (2006).
19. C. Palmes-Saloma and C. Blanca, unpublished results, ongoing project, *Development of Semiconductor Nanostructures as Optical Sensors to Quantify Anti-tumor Drug Efficacy in Animal Models,* OVCRD Open Grant, University of the Philippines – Diliman (2008).
20. G. Ranches, unpublished results, masteral thesis, Marine Science Institute, University of the Philippines – Diliman (2008).
21. E. Rodriguez, unpublished results, doctoral dissertation, Department of Mathematics, University of the Philippines – Diliman (2008).
22. D. Hanahan and R.A. Weinberg, *Cell* **100,** 57-70 (2000).
23. R.A. Weinberg, *The Biology of Cancer*, Garland Science, Taylor & Francis Group, LLC, New York, 2007, 796 pp.
24. N. López-Antón, A. Rudy, N. Barth, M.L. Schmitz, G.R. Pettit, K. Schulze-Osthoff, V.M. Dirsch, A.M. Vollmar, *J. Biol. Chem.* **281,** 33078-86 (2006).
25. R. David and H. Alla, in *Discrete, Continuous and Hybrid Petri Nets*, Springer-Verlag, 2005.
26. B.M. Olivera, J. Rivier, C. Clark, C.A. Ramilo, G.P. Corpuz, F.C. Abogadie, E.E. Mena, S.R. Woodward, D.R. Hillyard and L.J. Cruz. *Science* **249,** 257-63 (1990).

27. H. Terlau, K. Shon, M. Grilley, M. Stocker, W. Stühmer and B.M. Olivera, *Nature* **381,** 148-51 (1996).
28. B.M. Olivera, *Annu. Rev. Ecol. Syst.* **33,** 25-47 (2002).
29. B.M. Olivera, *J. Biol. Chem.* **281,** 31173-7 (2006).
30. B.M. Olivera, J.M. McIntosh, L.J. Cruz, F.A. Luque and W.R. Gray, *Biochemistry* **23,** 5087-90 (1984).
31. E.W. McCleskey, A.P. Fox, D. Feldman, L.J. Cruz, B.M. Olivera, R.W. Tsien and D. Yoshikami, *Proc. Nat. Acad. Sci. USA* **84,** 4327-31 (1987).
32. F.M. Heralde III, M. Watkins, J.P. Ownby, P.K. Bandyopadhyay, A.D. Santos, G.P. Concepcion and B.M. Olivera, *Nautilus* **121,** 131-138 (2007).
33. F.M. Heralde III, J. Imperial, P.K. Bandyopadhyayc, B.M. Olivera, G.P. Concepcion and A.D. Santos, *Toxicon* **51,** 890-897 (2007).
34. G.P. Concepcion and E.A. Padlan, *Med. Hypotheses* **60,** 699-701 (2003).
35. G.P. Concepcion, M.P.C. David and E.A. Padlan, *Med. Hypotheses* **64,** 919-924 (2004).
36. G.P. Concepcion and E.A. Padlan, *Med. Hypotheses* **65,** 865-867 (2005).
37. P.Y. Chou and G.D. Fasman, *Adv. Enzymol. Relat. Areas Mol. Biol.* **47,** 45-148 (1978).
38. L.J. McGuffin, K. Bryson and D.T. Jones, *Bioinformatics* **4,** 404-405 (2007).
39. M.P.C. David, J.J.T. Asprer, J.S.A. Ibana, G.P. Concepcion and E.A. Padlan, *Mol. Immunol.* **44,** 1342-1351 (2007).
40. C.A. Janeway, P. Travers, M. Walport and J.D. Capra, *Immunobiology: The Immune System in Health and Disease*, Fourth Edition, Garland Publishing, Taylor & Francis Group, New York (1999), 635 pp.
41. C. Branden and J. Tooze, *Introduction to Protein Structure*, Second Edition, Garland Publishing, Taylor & Francis Group, New York (1999), 410 pp.
42. E.A. Padlan, patent pending, USA (2006).
43. T. Horimito and Y. Kawaoka, *Nature Rev. Microbio.* **3,** 591-600 (2005).

Modeling Protein Conformations: Wiegel's Helical Polymers and the Hida-Streit White Noise Path Integral

Christopher C. Bernido*, M. Victoria Carpio-Bernido*, and Matthew George O. Escobido†,1

**Research Center for Theoretical Physics*
Central Visayan Institute Foundation
Jagna, Bohol 6308, Philippines
†Intel Solution Services ANZ
Intel Australia Pty Ltd, 111 Pacific Highway
North Sydney, 2060 Australia

Abstract. A stochastic model for biopolymers is discussed from a path integral approach to the solution of the Fokker-Planck equation. Explicit expressions for winding conformation probabilities are derived. The model incorporates experimentally observed features such as chirality, Brownian motion, the helix-turn-helix motif abundant in polypeptides, and overwinding of biopolymers when stretched up to a critical length.

Keywords: Biopolymers, Helix-turn-helix, winding probabilities, Brownian motion

PACS: 87.14.Gg, 87.15.He, 05.40.-a

INTRODUCTION

One of the striking features of a protein is its characteristic three-dimensional shape, or conformation, which allows it to perform a biological function. The manner in which a polypeptide folds, in fact, is dictated by the sequence of amino acids which make up its linear structure. What is not well understood, however, is how information encoded along the length of a polypeptide gives rise to a distinct three-dimensional shape. An example is the helix-turn-helix (HTH) motif of polypeptides which plays an important role in the molecular recognition and conformational adjustments of DNA-binding proteins. Consisting of about 20 amino acids, the HTH has two adjacent short α-helical segments separated by a turn. One of the α-helical segments is referrred to as the recognition helix, and the other a stabilization helix. In this paper, we use a stochastic model of biopolymers [1]-[3] and investigate the formation of an HTH motif, as well as varied features shown by experiments such as chirality, the stretching or unwinding of a polymer, and the observed overwinding of biopolymers when stretched [1], [4].

To guide a construction of an analytical stochastic model of biopolymers, we consider various experiments which reveal interesting features of the biomechanics of chain-like macromolecules. For instance, in experiments where beads (microspheres) are tethered by DNA molecules to study the stepwise bending of single DNA molecules [5], Brownian motion was observed using video microscopy. These and other experiments, e.g., snapshots in time of the random-walk path traveled by the center-of-mass of a labeled λ-DNA molecule [6], indicate that one can start to model polymer conformations using Brownian paths. This leads us to consider solutions of the Fokker-Planck equation. On the other hand, to handle the topological features of coiling, it is natural to

[1]Work done as visiting scientist at the Research Center for Theoretical Physics, Central Visayan Institute Foundation.

CP1021, *Stochastic and Quantum Dynamics of Biomolecular Systems,*
Proceedings of the 5th Jagna International Workshop, edited by C. C. Bernido and M. V. Carpio-Bernido

use the Feynman path integral to arrive at the probability density. We work with the basic idea of Edwards for entangled polymers [7]-[9], where the different possible conformations of a chain-like macromolecule are represented by paths, or trajectories, in the Feynman integral. We also make use of a subsequent result by Wiegel [10] which showed that a radial potential in this scenario can induce a helical winding of a polymer. With insights provided by experiments, we explicitly parametrize in this paper the Feynman paths in terms of the Brownian motion. The resulting probability function which solves the Fokker-Planck equation becomes an integral over a white noise functional.

Another important feature that needs to be incorporated in a model is the fact that biopolymers (e.g., proteins and DNA) exist in an aqueous environment and, hence, each segment or monomer in the chain-like molecule is in constant interaction with its environment. Since Brownian paths represent the biopolymer conformations in our model, we simulate the monomer-environment interaction by a length-dependent drift coefficient $f(s)$ in the Fokker-Planck equation where, $0 \leq s \leq L$ and L the total length of the biopolymer. The information encoded in the linear structure of the biopolymer, such as the sequence of amino acids and its interaction with the environment, is then made manifest by an s-dependent drift coefficient f. The path integral can be evaluated explicitly using white noise calculus [12], [13] and an explicit expression for the winding probabilities of a biopolymer is obtained. Several biomechanical properties of proteins can be exhibited by this model.

In the next section, we briefly review the path integral approach in solving the Fokker-Planck equation and the modeling of biopolymer conformations as Brownian paths in the path integral.

PATH INTEGRAL FOR POLYMER CONFORMATIONS

The solution of the Fokker-Planck equation for small intervals of time, $\tau << 1$, is given by the conditional probability density [14],

$$\begin{aligned} P(\mathbf{r}, t+\tau \mid \mathbf{r}_0, t) &= \left(\frac{1}{2\pi\tau}\right)^{3/2} \frac{1}{\sqrt{\det\left[\mathbb{B}(\mathbf{r}_0, t)\right]}} \\ &\times \exp\left\{-\frac{1}{2}\left[\left(\frac{\triangle\mathbf{r}}{\tau}\right) - \mathbf{A}(\mathbf{r}_0, t)\right]^T \mathbb{B}(\mathbf{r}_0, t)^{-1} \right. \\ &\left. \left[\left(\frac{\triangle\mathbf{r}}{\tau}\right) - \mathbf{A}(\mathbf{r}_0, t)\right]\tau\right\}. \end{aligned} \tag{1}$$

where, $\triangle\mathbf{r} = \mathbf{r} - \mathbf{r}_0$, $\mathbf{A}(\mathbf{r}_0, t)$ is the drift vector, and $\mathbb{B}(\mathbf{r}_0, t)$ the diffusion matrix. This short-time probability density may be used in the Chapman-Kolmogorov equation with conditional probabilities of intermediate events to give,

$$P(\mathbf{r}_1, t_1 \mid \mathbf{r}_0, t_0) = \lim_{N\to\infty} \int \prod_{j=1}^{N} P(\mathbf{r}_j, t_j \mid \mathbf{r}_{j-1}, t_{j-1}) \prod_{j=1}^{N-1} d\mathbf{r}_j. \tag{2}$$

Eq. (2) may symbolically be written as the path integral,

$$P(\mathbf{r}_1, t_1 \mid \mathbf{r}_0, t_0) = \int e^{-S(\mathbf{r})} \mathcal{D}[\mathbf{r}] \ , \tag{3}$$

with an effective action, $S(\mathbf{r}) = \int \mathcal{L}\, dt$, and a Lagrangian,

$$\mathcal{L} = \frac{1}{2}\left(\frac{d\mathbf{r}}{dt} - \mathbf{A}(\mathbf{r}, t)\right)^T \mathbb{B}(\mathbf{r}, t)^{-1} \left(\frac{d\mathbf{r}}{dt} - \mathbf{A}(\mathbf{r}, t)\right) . \tag{4}$$

Taking, $\mathbb{B} = D\mathbb{I}$, with $\mathbb{I}$ an identity matrix and D a constant diffusion coefficient, Eq. (1) becomes,

$$P(\mathbf{r}, t+\tau \mid \mathbf{r}_0, t) = \left(\frac{1}{2\pi D\tau}\right)^{3/2} \exp\left\{-\frac{1}{2D}\left[\left(\frac{\Delta \mathbf{r}}{\tau}\right) - \mathbf{A}\right]^2 \tau\right\}. \tag{5}$$

If the length travelled in time $\tau << 1$ is written as, $\triangle s = v\tau$, where v is the average speed of the random motion, Eq. (5) acquires the form,

$$P(\mathbf{r}, s_0 + \triangle s \mid \mathbf{r}_0, s_0) = \left(\frac{v}{2\pi D \triangle s}\right)^{3/2} \exp\left\{-\frac{v}{2D}\left[\left(\frac{\Delta \mathbf{r}}{\Delta s}\right) - \frac{\mathbf{A}}{v}\right]^2 \triangle s\right\}, \tag{6}$$

with, $\triangle s << 1$. We can define, $l = 3D/v$ which has a dimension of length, and the conditional probability density corresponding to Eq. (3) becomes,

$$P(\mathbf{r}_1, L \mid \mathbf{r}_0, 0) = \int \exp\left\{-\frac{3}{2l}\int_0^L \left[\frac{dr}{ds} - \frac{l}{3D}\mathbf{A}\right]^2 ds\right\} \mathcal{D}[\mathbf{r}], \tag{7}$$

where L is the total path length.

The path integral, Eq. (7), can be used to model a biopolymer where various paths of a Brownian motion with endpoints at $\mathbf{r}_0$ and $\mathbf{r}_1$ are viewed as possible conformations of a biopolymer. We take L to be the length of the chain-like macromolecule composed of N repeating units, each of length l, such that $L = Nl$. Each repeating unit of length l is allowed to rotate as a freely hinged rod to allow for bending and twisting. Since the integral in Eq. (7) is taken over all paths $\mathbf{r}(s)$ such that $\mathbf{r}(0) = \mathbf{r}_0$ and $\mathbf{r}(L) = \mathbf{r}_1$, with $0 \leq s \leq L$, the various conformations of the biopolymer can be investigated. This view of a polymer as a random walk consisting of N steps each of length l is also found in the early works of S. F. Edwards [7] and Prager and Frisch [8].

If we let $\mathbf{B}(s)$ be a Brownian motion, the paths can further be parametrized as,

$$\begin{aligned} \mathbf{r}(s) &= \mathbf{r}_0 + \kappa \mathbf{B}(s) \\ &= \mathbf{r}_0 + \kappa \int_0^s \boldsymbol{\omega}(\tau)\, d\tau, \end{aligned} \tag{8}$$

where $\boldsymbol{\omega}(s) = d\mathbf{B}/ds$ is a Gaussian random white noise variable [12] and κ a constant. This parametrization leads to a white noise functional as an integrand in Eq. (7) which will be discussed below.

HELICAL CONFORMATIONS

We now discuss situations where particular shapes of a biopolymer can arise. It would be helpful to recall that there exists monomer-solvent and monomer-monomer interactions which could be important in the determination of various conformations. For example, each amino acid in the sequence has side chains, or R groups, which could be nonpolar or hydrophobic, while some amino acids have polar or hydrophilic R groups, and still others either positively or negatively charged. These are sensitive to fluctuations in the environment.

How may an α-helix arise from the present stochastic model? To investigate helical structures, it is best to employ the circular cylindrical coordinates $\mathbf{r} = (\rho, \vartheta, z)$. In the paper by F. W. Wiegel [10] for a polymer represented by a random walk, he showed that a polymer subjected to a potential $V(\rho)$ winds around the z-axis at a distance $\rho = R$, where R is a minimum of the potential. This holds true for any interaction potential $V(\rho)$ at low temperatures. An example of this is, $V(\rho) = (C\rho^2) + (D/\rho^2)$, with $(C > 0,\ D > 0)$, which is repulsive at short distances and attractive at large distances. A close look at an α-helix does show that the amino acid side chains extend outward from the helical backbone. In other words, the R groups of each amino acid radiate to the outside of the helix [17]. For a rodlike α-helix oriented along the z-axis, the interaction of the side chains with the aqueous environment may contribute to a radial potential $V(\rho)$.

In view of the axial symmetry, we can simplify the study of the winding behavior of the biopolymer by projecting the random walk on the $\rho - \vartheta$ plane. One then considers a chain which lies on the plane perpendicular to the z-axis, with endpoints at $\boldsymbol{\rho}_0 = (\rho_0, \vartheta_0)$ and $\boldsymbol{\rho}_1 = (\rho_1, \vartheta_1)$. Given this scenario, the effect of a potential $V(\rho)$, with a minimum at $\rho = R$, is to induce a circular structure. In particular, Wiegel noted that the polymer conformations on the plane are confined to a narrow strip in the immediate vicinity of a circle around the origin with radius $\rho = R$. For this case we have, corresponding to Eq. (7), the probability function with $\boldsymbol{\rho} = (R, \vartheta)$,

$$P(\vartheta_1, \vartheta_0) = \int \exp\left[-\frac{1}{l}\int_0^L \left[R\left(\frac{d\vartheta}{ds}\right) - \frac{l}{2D} f(s)\right]^2 ds\right] \mathcal{D}[d\vartheta], \tag{9}$$

Here, the drift coefficient is taken to be length-dependent, i.e., $\mathbf{A} = f(s)$ to account for effects arising from interaction of the amino acid with the aqueous environment at each length segment. Corresponding to Eq. (8), the variable ϑ parametrized in terms of a Brownian motion B is given by,

$$\vartheta(L) = \vartheta_0 + \left(\sqrt{l}/R\right)\int_0^L \omega(s)\, ds, \tag{10}$$

where, $\omega(s) = dB/ds$, is a white noise variable. The parametrization, Eq. (10), leads to an integrand in Eq. (9) which is a white noise functional. The form of Eq. (9) differs from that previously treated in references [1]-[3]. For completeness, we present the essential steps of the analysis and its integration is discussed in the next section. This is followed by an examination of the winding probabilities using the variable ϑ which tracks the number of turns.

WHITE NOISE PATH INTEGRAL

The parametrization of ϑ in terms of a white noise variable $\omega(s)$, leads to the integrand of Eq. (9) of the form,

$$\exp\left[-\frac{R^2}{l}\int_0^L \left(\frac{d\vartheta}{ds}\right)^2 ds + \frac{R}{D}\int_0^L f(s)\left(\frac{d\vartheta}{ds}\right) ds - \frac{l}{4D^2}\int_0^L f(s)^2\, ds\right]$$

$$= \exp\left[-\int_0^L \omega(s)^2 ds + \frac{\sqrt{l}}{D}\int_0^L f(s)\ \omega(s) ds - \frac{l}{4D^2}\int_0^L f(s)^2\, ds\right]. \tag{11}$$

Since we are now dealing with a white noise functional, the path integration over $\mathcal{D}[d\vartheta] = d^\infty\vartheta$ becomes an integral over the Gaussian white noise measure defined in the infinite dimensional Hida calculus [12],[18], $d\mu(\omega) = N_\omega \exp[-\frac{1}{2}\int \omega(s)^2\, ds]\ d^\infty\omega$, where the exponential is responsible for the Gaussian fall-off and N_ω is a normalization factor[2]. In particular, the integral over $\mathcal{D}[d\vartheta]$ becomes an integral over, $N_\omega\, d^\infty\omega = \exp[\frac{1}{2}\int \omega(s)^2\, ds]\ d\mu(\omega)$. From the factor $\exp[\frac{1}{2}\int \omega(s)^2\, ds]$, and the exponential, $\exp\{-\int \omega(s)^2 ds\}$ on the right-hand-side of Eq. (11), we define the Gauss kernel I_0 given by,

$$I_0 = N \exp\left(-\frac{1}{2}\int_0^L \omega(s)^2\, ds\right), \tag{12}$$

[2] In quantum mechanics, Hida and Streit [11] used this procedure to express the Feynman integrand as a white noise functional.

where N is an appropriate normalization factor.

The path parametrization, Eq. (10) fixes the initial point ϑ_0, but the endpoint ϑ_1 of the Brownian paths still needs to be pinned down. To do this, we use the Donsker delta function [12], $\delta(\vartheta(L)-\vartheta_1)$. Furthermore, Eq. (9) deals with a Brownian motion confined to a circular topology. The Brownian paths can be classified topologically depending on the direction, clockwise or counterclockwise, and the number of times the polymer chain turns around the origin of the plane before ending at ϑ_1. These paths can be characterized by winding numbers [7,15,16] $n = 0, \pm 1, \pm 2, ...$ where, $n > 0$ signifies n turns counterclockwise around the origin; $n < 0$ means $|n|$ turns clockwise, and $n = 0$ signifies no winding[3]. To incorporate these possibilities, we use $\delta(\vartheta(L)-\vartheta_1+2\pi n)$, to reflect the topologically inequivalent trajectories that end at ϑ_1. This delta function summed over all possible n, together with Eqs. (11) and (12), allows us to write the probability function, Eq. (9) as,

$$\begin{aligned} P(\vartheta_1,\vartheta_0) &= \sum_{n=-\infty}^{n=+\infty} \int I_0 \exp\left[\frac{\sqrt{l}}{D}\int_0^L f(s)\ \omega(s)ds - \frac{l}{4D^2}\int_0^L f(s)^2\, ds\right] \\ &\quad \times \delta(\vartheta(L)-\vartheta_1+2\pi n)\ d\mu(\omega). \end{aligned} \tag{13}$$

The evaluation of Eq. (13) is facilitated by writing the Fourier representation of the delta function, i.e.,

$$\begin{aligned} P(\vartheta_1,\vartheta_0) &= \frac{1}{2\pi}\exp\left(-\frac{l}{4D^2}\int_0^L f(s)^2\, ds\right) \\ &\quad \times \sum_{n=-\infty}^{n=+\infty}\int d\lambda \exp\left[i\lambda\left(\vartheta_0-\vartheta_1+2\pi n\right)\right] \\ &\quad \times \int I_0 \exp\left\{ i\int_0^L \left[-\frac{i\sqrt{l}}{D}\int_0^L f(s) + \lambda\,\frac{\sqrt{l}}{R}\right]\omega(s)ds\right\}\ d\mu(\omega). \end{aligned} \tag{14}$$

The integration over $d\mu(\omega)$ is then just the T-transform of the Gauss kernel I_0 often used in white noise calculus [12]. The T-transform yields,

$$\begin{aligned} T(I_0)(\xi) &= \int I_0 \exp\left[i\int_0^L \omega(s)\xi(s)\, ds\right]\ d\mu(\omega) \\ &= \exp\left[-\frac{1}{4}\int_0^L \xi(s)^2\, ds\right]. \end{aligned} \tag{15}$$

where, $\xi(s) = \sqrt{l}\left[(-i/D)\, f(s)\ + (\lambda/R)\right]$. Eq. (14) acquires the form,

[3] A path moving clockwise, but no winding may be labelled by $n = 0^-$, and a path going counterclockwise with no winding by, $n = 0^+$.

$$P(\vartheta_1,\vartheta_0) = \frac{1}{2\pi}\sum_{n=-\infty}^{n=+\infty}\int d\lambda \exp\left(-\frac{lL}{4R^2}\lambda^2\right) \times \exp\left\{i\left[(\vartheta_0-\vartheta_1+2\pi n)+\frac{l}{2DR}\int_0^L f(s)\,ds\right]\lambda\right\}. \tag{16}$$

The remaining integral over λ may be evaluated in two ways. Employing the Poisson sum formula,

$$\frac{1}{2\pi}\sum_{n=-\infty}^{+\infty}\exp(in\phi) = \sum_{m=-\infty}^{+\infty}\delta(\phi+2\pi m), \tag{17}$$

Eq. (16) becomes,

$$P(\vartheta_1,\vartheta_0) = \frac{1}{2\pi}\sum_{m=-\infty}^{+\infty}\int \exp\left(-\frac{lL}{4R^2}\lambda^2\right) \times \exp\left\{i\left[(\vartheta_0-\vartheta_1)+\frac{l}{2DR}\int_0^L f(s)\,ds\right]\lambda\right\} \times\delta(\lambda+m)\;d\lambda, \tag{18}$$

and the integration is facilitated by the delta function to yield the result,

$$P(\vartheta_1,\vartheta_0) = \frac{1}{2\pi}\sum_{m=-\infty}^{+\infty}\exp\left(-\frac{lL}{4R^2}m^2\right) \times \exp\left\{-im\left[(\vartheta_0-\vartheta_1)+\frac{l}{2DR}\int_0^L f(s)\,ds\right]\right\}. \tag{19}$$

Alternatively, from Eq. (16), the integral over λ is evaluated as a Gaussian integral to yield,

$$P(\vartheta_1,\vartheta_0) = \sum_{n=-\infty}^{+\infty} P_n, \tag{20}$$

where,

$$P_n = \sqrt{\frac{R^2}{\pi lL}}\exp\left[-\frac{R^2}{lL}\left(\vartheta_0-\vartheta_1+2\pi n+\frac{l}{2DR}\int_0^L f(s)\,ds\right)^2\right]. \tag{21}$$

The P_n in Eq. (21) is the probability function for an n-times winding around the origin of the plane.

The probability that a polymer chain on the plane winds n-times around the z-axis is given by, $W(n,L) = P_n/P$. Using Eqs. (19) and (21) this yields,

$$W(n,L) = R\sqrt{\frac{4\pi}{Ll}}\frac{\exp\left[-\frac{R^2}{Ll}\left(2\pi n+\frac{l}{2DR}\int_0^L f\;ds\right)^2\right]}{\theta_3\left(\frac{l}{4DR}\int_0^L f\;ds\right)}, \tag{22}$$

where, for an arbitrary initial point we have set, $\vartheta_0 = \vartheta_1$. Here, the $\boldsymbol{\theta}_3(u)$ is the theta function [19],

$$\boldsymbol{\theta}_3(u) = 1 + 2\sum_{m=1}^{+\infty} \mathrm{q}^{m^2} \cos(2mu), \tag{23}$$

with $u = (l/4DR)\int f\ ds$, and $\mathrm{q} = \exp\left(-Ll/4R^2\right)$. Eq. (22) is an exact result obtained by evaluating Eq. (9).

If we designate a biopolymer which winds counterclockwise ($n > 0$) as left-handed, and the one which winds clockwise ($n < 0$) as right-handed, then the winding probability $W(n, L)$ will depend on the handedness, or chirality of the biopolymer [2]. From (22), the effect of chirality becomes evident since, $W(-n)/W(n) = \exp\left[(4\pi n/L)\int f\ ds\right]$. We note that when the drift coefficient $f(s)$ is zero, the winding probability $W(n, L)$, Eq. (22), becomes independent of chirality, or the sign of n.

DRIFT COEFFICIENT AS MODULATING FUNCTION

In the present stochastic model, we have allowed the drift coefficient $f(s)$ to be length-dependent where, $0 \leq s \leq L$, and L is the total length of the biopolymer. The drift coefficient $f(s)$, therefore, can carry the information of each monomer-environment interaction along the length of a biopolymer. Clearly, a choice of $f(s)$ can affect the winding probabilities as seen in Eq. (22). The $W(n, L)$ is an expression where linear information encoded along the length of a biopolymer can translate into a specific winding conformation for a polymer with $f(s)$ as a modulating function. The information, however, that is embodied by the possible available ways and variations of sequencing the 20 amino acids that make up the protein can be intractable. To cope with this difficulty, we expand the drift coefficient $f(s)$ in terms of a complete set of orthogonal functions. In particular, we may use, for example, the Laguerre polynomials and write,

$$f(s) = \sum_{m=0}^{\infty} k_m \alpha_m \mathbf{L}_m(\alpha_m s) \tag{24}$$

where k_m and α_m are constants, and $\mathbf{L}_m$ are the Laguerre polynomials. In Eq. (22), we have to evaluate, $\int f\ ds$, i.e.,

$$\int_0^L f\ ds = \sum_{m=0}^{\infty} k_m \alpha_m \int_0^L \mathbf{L}_m(\alpha_m s)\ ds. \tag{25}$$

The integral for each Laguerre polynomial can be done [19] and we obtain,

$$\int_0^L f\ ds = \sum_{m=0}^{\infty} k_m \left[\mathbf{L}_m(\alpha_m L) - \frac{1}{m+1}\mathbf{L}_{m+1}(\alpha_m L)\right]. \tag{26}$$

which could be used in Eq. (22) to get the probability for n-times winding of the form,

$$\begin{aligned} W(n, L) &= \left[\boldsymbol{\theta}_3\left(\frac{l}{4DR}\sum_{m=0}^{\infty} k_m\left[\mathbf{L}_m(\alpha_m L) - \frac{1}{m+1}\mathbf{L}_{m+1}(\alpha_m L)\right]\right)\right]^{-1} \\ &\quad \times R\sqrt{\frac{4\pi}{Ll}}\exp\left\{-\frac{R^2}{Ll}\left(2\pi n + \frac{l}{2DR}\right.\right. \\ &\quad \left.\left.\times \sum_{m=0}^{\infty} k_m\left[\mathbf{L}_m(\alpha_m L) - \frac{\mathbf{L}_{m+1}(\alpha_m L)}{m+1}\right]\right)^2\right\}. \end{aligned} \tag{27}$$

With Eq. (27) one can explore the various possible winding conformations by choosing specific values for the constants k_m and α_m. Moreover, for a very long chainlike biopolymer, $L = Nl >> 1$, the theta function $\boldsymbol{\theta}_3(u)$ in Eq. (27) approaches unity. We shall adopt this for simplicity in the succeeding sections.

HELIX-TURN-HELIX MOTIF

The helix-turn-helix structural motif may be demonstrated as a special case of Eq. (27). Let us consider the first four terms of the series, in particular, we let $k_m = k$ for $m = 0, ..., 3$, and $k_m = 0$ for higher values of m. We obtain from Eq. (27) the expression,

$$W(n,L) \approx R\sqrt{\frac{4\pi}{Ll}} \exp\left\{ -\frac{R^2}{Ll}\left(2\pi n + \frac{lk}{2DR} \times \sum_{m=0}^{3}\left[\mathbf{L}_m(\alpha_m L) - \frac{\mathbf{L}_{m+1}(\alpha_m L)}{m+1}\right]\right)^2\right\}. \quad (28)$$

Taking the values $R = 1$ nm, $l = 0.34$ nm, $\alpha_m = 1/\text{m}$, and $n = 40$, the graph of the winding probability $W(n, L)$ versus the biopolymer length L is given by Figure 1. The Graph with circles is for $(k/D) = 10^{4.5}$, and the solid line is for a much smaller value of the diffusion constant, i.e., $(k/D) = 10^6$.

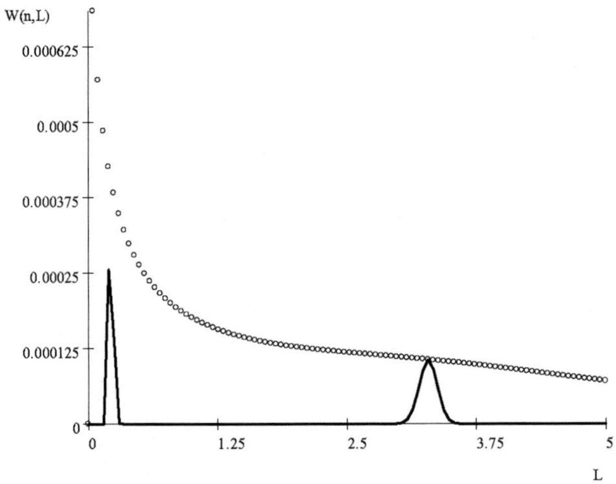

Figure 1. Circles for $(k/D) = 10^{4.5}$; Solid Line for $(k/D) = 10^6$

The maxima of $W(n, L)$, or the peaks of the solid line in the graph of Figure 1, show where winding is most probable. The two peaks, however, are separated by a section where winding is strongly inhibited. This situation suggests the appearance of a helix-turn-helix structural motif in polypeptides, where a helix is likely to be formed in each of the two peaks separated by a stretched portion.

OVERWINDING OF BIOPOLYMERS

Let us consider another feature that can be seen from Eq. (28). Experiments [4] have shown that when a biopolymer is stretched by forces up to 30 pN the strand overwinds until a critical length is reached [1]. This type of scenario may be seen from Eq. (28) by graphing the winding probability $W(n, L)$ for different values of L as shown in Figure 2, for $R = 1$ nm, $l = 0.34$ nm, and $(k/D) = 10^5$. The open box represents the probability for $n = 30$ winding, the dots for $n = 1,500$, and the solid line for $n = 3,000$. One observes that the $n = 30$ winding initially dominates for short L. However, as one extends a biopolymer and increases the value of L, higher values of the winding, e.g., $n = 1,500$ and $n = 3,000$ dominate, and the biopolymer is expected to overwind. The situation is reversed as one continues to extend the polymer where the $n = 30$ again becomes most probable. A critical value for the length L is reached where the $W(n, L)$ approaches a minimum, and further extensions lead to more probable smaller winding numbers (unwinding) of the macromolecular chain.

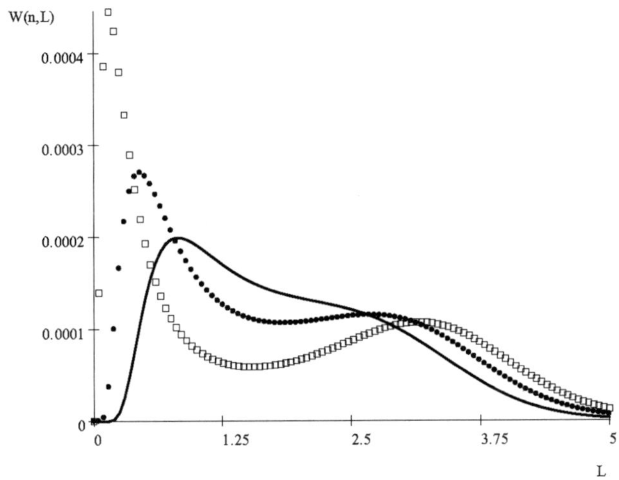

Figure 2. Open Box: $n = 30$; Dots: $n = 1500$; Solid line: $n = 3000$.

CONCLUSION

Using a path integral solution of the Fokker-Planck equation with a length-dependent drift coefficient $f(s)$, the possible conformations of a biopolymer are modelled by the various paths of a Brownian motion. In particular, the drift coefficient $f(s)$ is used to encode the monomer-environment interaction along the length of a biopolymer. The $f(s)$, for example, may simulate the biochemical information encoded in the amino acid sequence that make up a polypeptide. From the probability function which solves the Fokker-Planck equation, an analytic expression for the winding probabilities $W(n, L)$ of a polymer is obtained. The role of $f(s)$ as a modulating function for the winding conformations is then demonstrated by expanding it in terms of Laguerre polynomials. This allowed us to demonstrate a winding probability which indicates a helix-turn-helix structural motif. Other experiment-based features of the model are also discussed, such as chirality and the overwinding of biopolymers. Extensions of this work will consider other forms of the drift coefficient $f(s)$, and corresponding winding probability $W(n, L)$, for comparison with experiments, and further refinement of the model.

REFERENCES

1. C. C. Bernido and M. V. Carpio-Bernido, *Phys. Lett.* A**369**, 1-4 (2007).

2. C. C. Bernido, M. V. Carpio-Bernido, and J. B. Bornales, *Phys. Lett.* A**339**, 232-236 (2005).

3. C. C. Bernido and M. V. Carpio-Bernido, *J. Phys. A: Math. Gen.* **36,** 4247-4257 (2003).

4. T. Lionnet, S. Joubaud, R. Lavery, D. Bensimon, and V. Croquette, *Phys. Rev. Lett.* **96**, 178102 (2006); J. Gore, Z. Bryant, M. Nöllmann, M. U. Le, N. R. Cozzarelli, and C. Bustamante, *Nature* **442**, 836-839 (2006).

5. See, e.g., S. F. Tolić-Norrelykke, M. B. Rasmussen, F. S. Pavone, K. Berg-Sorensen, L. B. Oddershede, *Biophys. J.* **90,** 3694-3703 (2006); Y. C. Sasaki et al, *Phys. Rev. Lett.* **87**, 248102 (2001).

6. D. E. Smith, T. T. Perkins, and S. Chu, *Phys. Rev. Lett.* **75**, 178102 (1995).

7. S. F. Edwards, *Proc. Phys. Soc. London,* **91**, 513-519 (1967).

8. S. Prager and H. L. Frisch, *J. Chem. Phys.* **46**, 1475 (1967).

9. A. Inomata and V. A. Singh, *J. Math. Phys.* **19**, 2318-2323 (1978).

10. F. W. Wiegel, *J. Chem. Phys.* **67**, 469-472 (1977).

11. L. Streit and T. Hida, *Stoch. Proc. Appl.* **16**, 55-69 (1983).

12. T. Hida, H.-H. Kuo, J. Potthoff, and L. Streit, *White Noise. An Infinite Dimensional Calculus* (Kluwer, Dordrecht, 1993); N. Obata, *White Noise Calculus and Fock Space*, Lecture Notes in Mathematics, Vol. 1577 (Springer, Berlin, 1994); H.-H. Kuo, *White Noise Distribution Theory* (CRC, Boca Raton, FL, 1996).

13. See, also, C. C. Bernido and M. V. Carpio-Bernido, *J. Math. Phys.* **43**, 1728-1736 (2002).

14. See, also, H. Risken, *The Fokker-Planck Equation* (Springer-Verlag, Berlin, 1996).

15. L. S. Schulman, *Techniques and Applications of Path Integration* (Wiley, New York, 1981).

16. M. G. Laidlaw and C. M. DeWitt, *Phys. Rev.* D**3**, 1375 (1971).

17. See, e.g., M. Singer and P. Berg, *Genes and Genomes* (Univ. Science Books, Mill Valley, 1991).

18. M. de Faria, J. Potthoff, and L. Streit, *J. Math. Phys.* **32**, 2123-2127 (1991).

19. I. S. Gradshteyn and I. M. Ryzhik, *Table of Integrals, Series, and Products*, 5th ed. (Academic Press, San Diego, 1994).

Winding probabilities á la Wiegel with lineally dependent potentials

Mark Nolan P. Confesor and Jinky B. Bornales

Department of Physics, Mindanao State University - Iligan Institute of Technology, Andres Bonifacio Ave., Tibanga, Iligan City, 9200 Philippines

Abstract. The probability for a polymer to wind n-times around an obstacle for a class of lineally dependent potentials were calculated using the differential approach introduced by F.W.Wiegel. The results agree with that obtained by C.C.Bernido and Ma.V.Carpio-Bernido using white noise analysis approach.

Keywords: constrained polymer conformations, winding probabilities
PACS: 87.15.ad, 87.15.La, 36.20.Ey

INTRODUCTION

Advances in the techniques of molecular experiments has led to the unraveling of biological macromolecules in their native and entangled states [1]. The occurrence of these macromolecules in these states suggest that topological constraints play a crucial role in the operations of biomolecular systems. The DNA with its α-helix conformation is a good example of a biomolecular system with topological constraint. The importance of the constraint can be seen on how the DNA would coil and uncoil to pass genetic information. Moreover it is remarkable that DNA possess sequence-dependent curvature which arises chirality [2]. Also for circular DNA it is believe that certain tertiary structural properties may have been summarized into the DNA sequenze during evolution [3].

In this article, we examine the coiling of a macromolecule on an obstacle where we incorporate sequenze dependence of the tertiary stucture of the macromolecule by a lineally dependent potential [4].

POLYMER ENTANGLEMENT

In the simple entanglement, a polymer's motion is constrained by a straight line obstacle at the origin. The monomers of the polymer interacts with the obstacle by a potential V_{ext} and the monomer-monomer interaction is given by a potential V_{mm}. To distinguish the different possible windings, we introduced the winding number n, such that $n = 0, \pm 1, \pm 2, \ldots$ and $n \geq 0$ means n turns counterclockwise and $n \leq -1$ means $|n+1|$ clockwise turns on the obstacle.

CP1021, *Stochastic and Quantum Dynamics of Biomolecular Systems, Proceedings of the 5th Jagna International Workshop,* edited by C. C. Bernido and M. V. Carpio-Bernido

The conformational sum for an n times entanglement is given, in three dimensions by [5]

$$Q_n(r_N,N|r_0,0)=\int_{r_0}^{r_N}\exp\left(-\frac{3}{2l^2}\int_0^N\left(\frac{d\vec{r}}{dv}\right)^2dv-\int_0^N Vdv\right.$$
$$\left.-\int_0^N\int_0^N V_{mm}dv\,dv'\right)\delta\left(n-L\left[\vec{r}(v)\right]\right)d\left[\vec{r}(v)\right] \quad (1)$$

for a polymer whose endpoints are located at $\vec{r}_0$ and $\vec{r}_N$. The *Kuhn segment length* l is the fixed distance between two monomer and v is an ordinal number which runs from zero to N. The Dirac delta function in Eq.(1) fixes the total number of windings around the obstacle. The functional $L\left[\vec{r}(v)\right]$ has the form $L\left[\vec{r}(v)\right]=\int\vec{F}_t\cdot d\vec{r}$ for a straight line obstacle where $\vec{F}_t=\frac{1}{2\pi R}\hat{\theta}$ and R being the radius of the winding which we take it as constant. The winding probability is then defined as the ratio

$$W_n(N)=\frac{Q_n(N)}{Q(N)} \quad (2)$$

where $Q(N)$ is the sum for all possible conformations given by $Q(N)=\sum_n Q_n(N)$.

ENTANGLEMENT WITH LINEALLY DEPENDENT POTENTIAL

Following the work of C.C.Bernido and M.V. Carpio-Bernido [4], we simulate the effect of the bases on the structure of the DNA by choosing the potentials to be lineally dependent and of the form $V=\frac{d\tilde{f}}{dv}\theta(v)$. This potential carries with it the net effect of both the external and monomer-monomer interaction. We approximate the form of $\tilde{f}$ from the monomer-monomer interaction terms of Eq.(1) and have $\tilde{f}=\frac{1}{N}\int_0^N f(v)dv$.

The potential $V=\frac{d\tilde{f}}{dv}\theta$ can be cast in the following form,

$$\int_0^N\frac{d\tilde{f}}{dv}\theta dv=\left[\tilde{f}_N\theta_N-\tilde{f}_0\theta_0\right]-\int_0^N\tilde{f}\frac{d\theta}{dv}dv. \quad (3)$$

Plugging this to Eq.(1) we get,

$$Q_n(r_N|r_0) = \int_{-\infty}^{\infty}\exp\left[\tilde{f}_N\theta_N-\tilde{f}_0\theta_0\right]Q_k\exp\{2\pi kni\}dk, \quad (4)$$

where Q_k is

$$Q_k=\int_{r_0}^{r_N}\exp\left(-\frac{3}{2l^2}\int_0^N\left(\frac{d\vec{r}}{dv}\right)^2dv-2\pi ki\int_0^N\vec{B}\cdot\frac{d\vec{r}}{dv}dv\right)d\left[\vec{r}(v)\right]. \quad (5)$$

and $\vec{B}=\vec{F}_t+\frac{i\vec{\tilde{F}}}{2\pi k}$. Path-integral of the type given in Eq.(5) has been extensively studied by F.W.Wiegel[5] and was found to satisfy the following diffusion like equation,

$$\left[\frac{\partial}{\partial N}-\frac{l^2}{6}\nabla^2-ik\Omega\frac{l^2}{3R^2}\frac{\partial}{\partial\theta}+\frac{l^2}{6}k^2\frac{\Omega^2}{R^2}\right]Q_k=\delta\left(\vec{r}_N-\vec{r}_0\right)\delta(N) \quad (6)$$

where $B^2 = \frac{\Omega^2}{4\pi^2 R^2}$ and $\Omega^2 = \left(1+\frac{\tilde{f}}{k}i\right)^2$.

We then solve for Q_k by power series expansion

$$Q_k = \sum_a f_a(\vec{r}_N) f_a^*(\vec{r}_0)\ \exp(-\lambda(a)N) \tag{7}$$

where the eigenvalues λ_a satisfy

$$\left[-\frac{l^2}{6}\nabla^2 - ik\Omega\frac{l^2}{3R^2}\frac{\partial}{\partial\theta} + \frac{l^2}{6}k^2\frac{\Omega^2}{R^2}\right] f_a = (\lambda_a) f_a. \tag{8}$$

To solve for f_a, we employ separation of variables and impose appropriate boundary conditions. The conformational sum for an n-times entanglement, Eq.(4), is then found to be

$$\begin{aligned} Q_n(N) &= C\exp\left\{\frac{Nl^2}{6R^2}\tilde{f}^2\right\}\int_{-\infty}^{\infty}\exp\left\{-\frac{Nl^2}{6R^2}k^2 + \left(2\pi n - \frac{Nl^2}{3R^2}\tilde{f}\right)ik\right\}dk \\ &= \frac{R}{l}\sqrt{\frac{6\pi}{N}}\ \exp\left\{\frac{Nl^2}{6R^2}\tilde{f}^2\right\}\ \exp\left\{2\pi n\tilde{f} - \frac{6\pi^2 n^2 R^2}{Nl^2} - \frac{Nl^2\tilde{f}^2}{6R^2}\right\}. \end{aligned} \tag{9}$$

where C is just $\sum_p \exp\left\{-\frac{Nl^2}{6R^2}\left(p^2R^2\right)\right\}$. Summing Eq.(9) for all possible values of n, we arrive at the form of $Q(N)$, to be

$$Q(N) = C\exp\left\{\frac{Nl^2}{6R^2}\tilde{f}^2\right\}\sum_{m=-\infty}^{\infty}\exp\left\{-\frac{Nl^2}{6R^2}m^2 + \left(\frac{Nl^2}{3R^2}\tilde{f}\right)im\right\}, \tag{10}$$

where we have used the Poisson sum formula ($\sum_{n=-\infty}^{\infty}\exp\{2\pi nki\} = \sum_{m=-\infty}^{\infty}\delta(k+m)$).

The winding probability is just the ratio of $Q(N)$ and $Q_n(N)$, that is

$$W_n(N) = \frac{R}{l}\sqrt{\frac{6\pi}{N}}\frac{\exp\left\{2\pi n\tilde{f} - \frac{6\pi^2n^2R^2}{Nl^2} - \frac{Nl^2\tilde{f}^2}{6R^2}\right\}}{\theta_3\left(\frac{Nl^2}{6R^2}\tilde{f}\right)} \tag{11}$$

where we write the denominator in terms of the theta function $\theta_3(u)$ [7]. In two dimensions, we perform the substitution $l^2 \to \frac{3}{2}l^2$ [5] to obtain

$$W_n(N) = \frac{R}{l}\sqrt{\frac{4\pi}{N}}\frac{\exp\left\{-\frac{R^2}{Nl^2}\left(2\pi n - \frac{l}{2R^2}\int_0^{Nl} f(s)\,ds\right)^2\right\}}{\theta_3\left(\frac{l}{4R^2}\int_0^{Nl} f(s)\,ds\right)}, \tag{12}$$

where $0 \le s(=vl) \le L$, L being the contour length of the polymer which is equal to Nl. For a very long polymer the denominator approaches to unity and Eq.(12) becomes

$$W_n(N) \approx \frac{R}{l}\sqrt{\frac{4\pi}{N}}\ \exp\left\{-\frac{R^2}{Nl^2}\left(2\pi n - \frac{l}{2R^2}\int_0^L f(s)\,ds\right)^2\right\}. \tag{13}$$

Eq.(13) is similar in form with the expression obtained by C.C. Bernido and Ma.V. Carpio-Bernido [4] where they have used the white noise functional approach.

With Eq.(13), we can then consider various forms of $f(s)$ that would exhibit some of the observed features of an entangled DNA. Among the forms of $f(s)$ that have been considered so far are [4]

1. $f(s) = ks^p$, where k is a constant and p is an integer,
2. $f(s) = k \cos(\alpha s)$, where k and α are constants and
3. $f(s) = k\,\alpha \mathbf{L}_2(\alpha s)$, where $\mathbf{L}_2$ is the Laguerre polynomial and k and α are constants.

ACKNOWLEDGMENTS

We would like to thank Dr. Christopher C. Bernido and Dr. Ma. Victoria Carpio-Bernido for giving us the opportunity to participate in the 2005 and 2007 Jagna Conferences. We are also equally indebted to Prof. Frederick Wiegel for the helpful discussions.

REFERENCES

1. F.Ritort, *J.Phys.:Condens.Matter*, **18** R531-R583 (2006).
2. P.J. Hagerman, *Annu. Rev. Biochem.*, **59** 755-781 (2000).
3. A.G. Pedersen, L.J. Jensen, S. Brunak, H.H. Staerfeldt, and D.W. Ussrey, *J. Mol. Biol.* **299**, 907-930 (2000).
4. C.C. Bernido, and M.V.Carpio Bernido, *J. Phys. A:Math.Gen.* **36**, 4247-4257 (2003).
   C.C. Bernido, M.V.Carpio Bernido, and J.B. Bornales, *Phys. Letters A* **339**, 257-266 (2005).
   C.C. Bernido, and M.V. Carpio-Bernido, *Physics Letter* **369**, 1-4.s (2007)
5. F. W. Wiegel, *J. Chem. Phys.* **67** 469-72 (1977).
   F.W. Wiegel, *Introduction to Path-Integral Methods in Physics and Polymer Science*, World Scientific Publishing Co Pte Ltd., 1986.
   F. W. Wiegel, *Phase Transitions and Critical Phenomena,* edited by Domb and Lebowitz, Academic Press, 1983, pp. 101-149.
6. S.F. Edwards, *Proc.Phys.Soc.*, **91** 513-519 (1967).
   S.Prager and H.L.Frisch, *J. Chem. Phys.*, **46** 1475-1483 (1967).
   N. Saito and Y. Chen, *J. Chem. Phys.*, **46** 1475-1483 (1973).
7. I. S. Gradshteyn and I. M. Ryzhik, *Table of Integrals, Series, and Products, 5th edtn.*, Academic Press, Inc., 1994.

Evaluation of the Probability Distribution for the Area Enclosed by a Polymer Loop Using White Noise Analysis

Beverly V. Gemao and Jinky B. Bornales

Physics Department, Mindanao State University-Iligan Institute of Technology, Andres Bonifacio Ave., Iligan City 9200, Philippines

Abstract. The probability density that a polymer loop conformation encloses an area A is calculated using white noise functional approach. In this approach, the paths of the polymer conformation, $x(v)$ and $y(v)$, are parametrized in terms of Brownian motion. Evaluation of the path integral describing the system is then facilitated using the T- transform in white noise analysis.

Keywords: polymer conformation, probability distribution and white noise analysis
PACS: 01.30.Cc, 02.30.Sa, 36.20.Ey

INTRODUCTION

A polymer physically constrained by a concentrated polymer solution was modelled by Brereton and Butler as a random walk which encloses a fixed area [1]. This representation was adopted by Khandekar and Wiegel in investigating the statistical mechanical properties of the system, one of which is the probability density that a polymer conformation will enclose an algebraic area A [2]. In their evaluation, they used a straightforward calculation of the path integral describing the system.

In this paper we present another way of solving the probability density, i.e., by applying white noise analysis.

THE MODEL

In this section we briefly discuss the model of the problem as presented in [2]. The probability density of a free, two-dimensional random flight whose two ends are fixed at $\mathbf{r}_0$ and $\mathbf{r}_1$ has the Wiener integral representation[2, 3]

$$P(\mathbf{r}_1,N|\mathbf{r}_0,0) = \int_{\mathbf{r}_0}^{\mathbf{r}_1} \exp\left[-\frac{1}{l^2}\int_0^N \left(\frac{dr}{dv}\right)^2 dv\right] d[\mathbf{r}(v)], \tag{1}$$

where for non-interacting polymer, l is the distance between two monomers and v is the number of monomers with increasing value from 1 to N. While the probability distribution for the area A enclosed by a polymer loop is given by the path integral

$$P(A,N) = N_0 \int_{\mathbf{r}_0}^{\mathbf{r}_1} \delta\left[A - \int_0^N \frac{1}{2}\left(x\frac{dy}{dv} - y\frac{dx}{dv}\right)dv\right] \exp\left\{-\frac{1}{l^2}\int_0^N \left(\frac{d\mathbf{r}}{dv}\right)^2 dv\right\} d[\mathbf{r}(v)], \tag{2}$$

CP1021, *Stochastic and Quantum Dynamics of Biomolecular Systems, Proceedings of the 5th Jagna International Workshop,* edited by C. C. Bernido and M. V. Carpio-Bernido

where N_0 is the normalization constant and

$$A = \frac{1}{2}\int_0^N \left(x\frac{dy}{dv} - y\frac{dx}{dv}\right) dv \tag{3}$$

is the algebraic area A which is enclosed by a given closed polymer conformation $\mathbf{r}(v)$. The sign convention used is that A is counted positive(negative) if the enclosed area is located to the left(right) of the curve when traced by increasing value of v.

Rewriting the Dirac delta function as a Fourier integral, the probability density, Eq.(2), can be written as

$$P(A,N) = \frac{N_0}{2\pi}\int_{-\infty}^{\infty} \tilde{P}(g,N)\exp(igA)dg \tag{4}$$

where

$$\tilde{P}(g,N) = \int_{\mathbf{r}_0}^{\mathbf{r}_1} \exp\left(-\int_0^N \tilde{L}dv\right) d[\mathbf{r}(v)] \tag{5}$$

with

$$\tilde{L} = \frac{1}{l^2}\left\{\left(\frac{dx}{dv}\right)^2 + \left(\frac{dy}{dv}\right)^2\right\} + \frac{1}{2}ig\left(x\frac{dy}{dv} - y\frac{dx}{dv}\right). \tag{6}$$

We shall evaluate Eq.(5) in the framework of white noise analysis. The evaluation will be presented in the next section.

EVALUATION OF THE PROBABILITY DISTRIBUTION IN THE FRAMEWORK OF WHITE NOISE ANALYSIS

To employ white noise analysis in the evaluation of Eq. (5), we parametrize the paths as

$$x(L) = x_0 + lB_x(L) \tag{7}$$
$$y(L) = y_0 + lB_y(L), \tag{8}$$

where $B_x = \int_0^N \omega_x dv$ and $B_y = \int_0^N \omega_y dv$ are Brownian motions parametrized by v, $0 \leq v \leq N$, ω is the white noise variable, and L is the length of the polymer loop. With the parametrizations, the integral over the measure $d[\mathbf{r}(v)]$ now becomes an integral over the Gaussian white noise measure $d\mu(\omega)$, that is [4]

$$d[\mathbf{r}(v)] \rightarrow N_{xy} \exp\left[\frac{1}{2}\int_0^N ({\omega_x}^2 + {\omega_y}^2)dv\right] d\mu(\omega_x)d\mu(\omega_y). \tag{9}$$

Furthermore, we need to fix the endpoints and this can be done using the Donsker delta functions. We further note that we are considering a closed polymer conformation, i.e.,$y_1 = y_0$ and $x_1 = x_0$, and so we can write the Donsker delta functions as

$$\delta(x(L) - x_1) = \delta(x_0 + lB_x - x_1) = \delta(lB_x), \tag{10}$$
$$\delta(y(L) - y_1) = \delta(y_0 + lB_y - y_1) = \delta(lB_y). \tag{11}$$

Expressing the Donsker delta functions, Eqs. (10) and (11), in their Fourier representations and incorporating Eq. (7-11) to Eq. (5), we then have

$$\tilde{P}(g,N) = \frac{N_{xy}}{4\pi^2}\int_{-\infty,p_y}^{\infty}\int_{-\infty,p_x}^{\infty}\int_{\mathbf{r}_0}^{\mathbf{r}_1} \exp\Big\{\int_0^N \Big[-\frac{1}{2}(\omega_x{}^2+\omega_y{}^2)+\frac{\lambda l}{2}(x_0\omega_y - y_0\omega_x)$$
$$+\ \lambda l^2\Big[\frac{1}{2}(B_x dB_y - B_y dB_x)\Big] + ip_x l\omega_x + ip_y l\omega_y\Big]\Big\} d\mu(\omega_x) d\mu(\omega_y) dp_x dp_y, \quad (12)$$

where $\lambda = -ig$. We note that $\int \frac{1}{2}(B_x dB_y - B_y dB_x)$ is defined as the Levy's stochastic area where the evaluation of this can be done in the probability space of one-dimensional white noise as follows [5]:

$$B_x(v) = \int_o^v \omega(v')dv' \quad (13)$$
$$B_y(v) = \int_{-v}^0 \omega(v')dv'. \quad (14)$$

With Eqs.(13-14) and using the expression

$$\frac{d}{db}\int_b^a f(x)dx = -f(b), \quad (15)$$

Eq. (12) becomes

$$\tilde{P}(g,N) = \frac{N_{xy}}{4\pi^2}\int_{-\infty,p_y}^{\infty}\int_{-\infty,p_x}^{\infty}\int_{\mathbf{r}_0}^{\mathbf{r}_1} \exp\Big\{-\int_N^{-N}\frac{1}{2}\omega(v)^2 dv$$
$$+\int_{-N}^0 \Big[ip_y l\omega_v + \frac{1}{2}\lambda l x_0\omega_v\Big]dv + \int_0^N \Big[ip_x l\omega_v - \frac{1}{2}\lambda l y_0\omega_v\Big]dv$$
$$+\ \lambda l^2 \int_{R^2}\omega(v)F_s(v,v')\omega(v')dv\,dv'\Big\} d\mu(\omega)dp_x dp_y, \quad (16)$$

where

$$F_s(v,v') = \frac{1}{4}\Big[X^{(v)}_{[-N,0]}X^{(v')}_{[0,-v]} + X^{(v')}_{[-N,0]}X^{(v')}_{[0,-v]} + X^{(v)}_{[0,N]}X^{(v')}_{[-v,0]} + X^{(v')}_{[0,N]}X^{(v)}_{[-v',0]}\Big] \quad (17)$$

with $X^{(c)}_{[a,b]}$ as the integration over c from a to b.

We note that the integration over $d\mu(\omega)$ in Eq.(16) is just the T-transform of the functional $\Phi(\omega)$, that is [4]

$$T\Phi(\omega) = \int \exp(i\langle\omega,\xi\rangle)\ \Phi(\omega)d\mu(\omega) \quad (18)$$

where from Eq.(16) we identify $\Phi(\omega)$ to be

$$\Phi(\omega) = \exp\Big\{-\frac{1}{2}\langle\omega,\omega\rangle + \lambda l^2\langle\omega, F_s\omega\rangle\Big\} \quad (19)$$

with

$$\xi = \left(p_y l \; - \; \frac{1}{2} i\lambda l x_0\right) X_{[-N,0]} + \left(p_x l + \frac{1}{2} i\lambda l Y_0\right) X_{[0,N]}. \tag{20}$$

Calculations of the T-transform of Φ yields

$$\begin{aligned} T\Phi(\omega) &= \left[\cos\frac{\lambda l^2 N}{4}\right]^{-1} \exp\left\{ -\frac{\pi}{\lambda l^2} \tan\left(\frac{\lambda l^2 N}{4}\right) \left[p_y^2 l^2 \right.\right. \\ &- \left.\left. i\lambda l^2 x_0 p_y - \frac{\lambda^2 l^2 x_0^2}{4} + p_x^2 l^2 - i\lambda l^2 x_0 p_x - \frac{\lambda^2 l^2 y_0^2}{4}\right]\right\}. \end{aligned} \tag{21}$$

We are left with the integrations over p_x and p_y and this can be done using the one-dimensional Gaussian integral formula which gives us

$$\tilde{P}(g,N) = \frac{N_0 \lambda}{4\pi \sin\left(\frac{\lambda l^2 N}{4}\right)}. \tag{22}$$

Eq.(22) agrees with the result obtained by Wiegel and Khandekar[3]. With Eq.(22),we are left with the integration over dg to obtain $P(A,N)$. Integration over dg yields

$$P(A,N) = \left[2Nl^2 \cosh^2\left(\frac{2\pi A}{Nl^2}\right)\right]^{-1}. \tag{23}$$

CONCLUSION

In this paper we have obtained the expression for the probability distribution for the area enclosed by a polymer loop using white noise analysis. This problem would serve as a stepping stone to a system with the same constraints but with the presence of some potentials.

ACKNOWLEDGMENTS

BVGemao wishes to acknowledge Dr. C.C. Bernido and Dr. M.V. Carpio-Bernido for their series of lectures on white noise analysis in our department.

REFERENCES

1. M.G. Brereton, and C. Butler *J.Phys. A*, **20**(1987)3955
2. D.C. Khandekar, and F.W. Wiegel, *J. Stat. Phys.*, **53**(1988)1073-1080
3. F.W. Wiegel, and D.C. Khandekar, *Int. J. Mod. Phys. B*, **6**(1992) 3147-3154
4. C.C. Bernido, and M.V. Carpio-Bernido, *J. Phys A:Math. Gen.* **36**(2003)4247-4257
5. D. de Falco and D.C. Khandekar, *Stoch. Proc., Appl.* **29**(1988)257-266

Single-cell growth analysis in a mixed cell culture

Jun Ando[a,b], Mary Grace P. Bato[a] and Vincent Ricardo Daria[a]

[a] *National Institute of Physics, University of the Philippines, Quezon City, 1101 Philippines*
[b] *Department of Applied Physics, School of Engineering, Osaka University, 2-1 Yamadaoka, Suita, Osaka 565-0871, Japan*

Abstract. We perform single cell analysis of cell growth in a mixed cell culture. Two species of yeast cells: *Saccharomyces cerevisiae* and *Candida albicans*, are optically trapped using focused continuous-wave near infrared laser. Cell growth for both cells is inhibited only when the two species of cells are in contact with each other. This indicates cell-cell interaction mediated cell growth inhibition mechanism. Single cell level analysis of cell growth studied here contributes to the further understanding of yeast growth arrest in a mixed yeast culture.

Keywords: Growth inhibition, single cell level analysis, optical trapping.
PACS: S 87.64.-t

INTRODUCTION

Yeast cells are one of the most important organisms in wine making as it determines its flavor and taste. Among the many kinds of yeast strains, *Saccharomyces cerevisiae (S cerevisiae)* are mainly used for this process, because it has a strong resistance to ethanol. However, for deeper and more complex taste of wine production, other strains of yeast cells are mix-cultured with *S cerevisiae.* In a mixed culture, there is an increase in induced interactions between these different strains, resulting in early death of the other strains before *S. cerevisiae.* Some researchers associate early death to the difference of ethanol tolerance between *S cerevisiae* and the other species. Other growth inhibiting factors such as nutrient limitation, killer toxins [1], and farnesol [2], cell-to-cell contact [3,4] and confinement stress [5] are also attributed to this phenomenon. However, almost all these research activities are limited to macroscopic observation and bulk analysis. Further verification of growth inhibition mechanism is needed to perform single cell level analysis.

Laser based optical trapping (optical tweezers) is well known as a contact free (sterile) and a non-invasive tool for the cells and manipulation of organelles. Tightly focused laser beam induces gradient forces, which confines biological sample towards the laser focus. Since Ashkin *et al.* reported the first laser-based single cell and organelle manipulation [6,7], optical tweezers is now used for wide field of biological research, such as measurement of kinesin molecules movement [8], drug screening [9], cell sorting [10], and investigation of DNA mechanics [11].

CP1021, *Stochastic and Quantum Dynamics of Biomolecular Systems, Proceedings of the 5th Jagna International Workshop,* edited by C. C. Bernido and M. V. Carpio-Bernido

In this paper, we use dynamic microscopy by using optical tweezers for single cell level analysis of *S cerevisiae* in a mixed culture with *Candida albicans* (*C albicans*) using a continuous wave near infrared laser. By controlling the position of the cells via an optical trap, we show that physical contact between the two species of yeasts shows growth inhibition for both species.

MATERIAL AND METHODS

The schematic of the experimental set-up for optical trapping of yeast cells is shown in Fig.1. We used a Titanium:Sapphire (Ti:S) laser (Tsunami, Spectra Physics) pumped with a $Nd:YV0_4$ (Millenia V4, Spectra Physics). In this experiment, the laser is operated in continuous-wave (CW) mode. The wavelength is centered at 780 nm. The laser is introduced to a custom-built inverted laser-scanning microscope, and brought to focus inside sample using a water immersion objective lens (60×/NA 1.20, Olympus). The microscope includes a dichroic mirror to direct the near infrared laser to the objective lens while allowing images of cells at the focus of the same objective lens to be viewed on a CCD camera. The laser focus can be scanned along the x-y plane using two galvano-mirrors. Time-lapse sequences of bright field images are recorded. Image acquisition and control of the scanning-mirrors are achieved using a computer. The software to control the instrumentation is written in LabView (National Instruments).

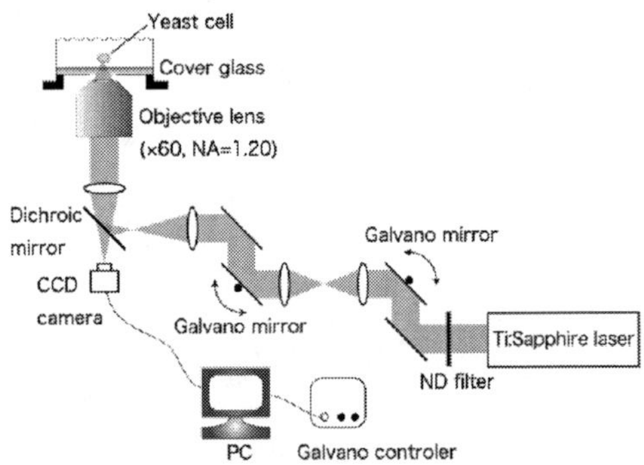

FIGURE 1. Experimental setup for laser trapping of living yeast cells

RESULTS AND DISCUSSION

To verify cell viability and growth under a laser trap, we observed the growth of an optically trapped yeast cell. *S cerevisiae* is known to grow via the division process of budding. Fig. 2 shows time-lapse bright field images of optically trapped yeast cell growth using CW mode Ti:S laser at 10 mW of laser power measured at the laser focus. Black triangles indicate the laser focus, and arrows indicate yeast budding. Because of the trapping force, the yeast cell stands perpendicular to the glass substrate, as shown in Fig.2 (20min to 80min). Even under the Ti:S laser irradiation in CW mode, yeast cell shows budding growth.

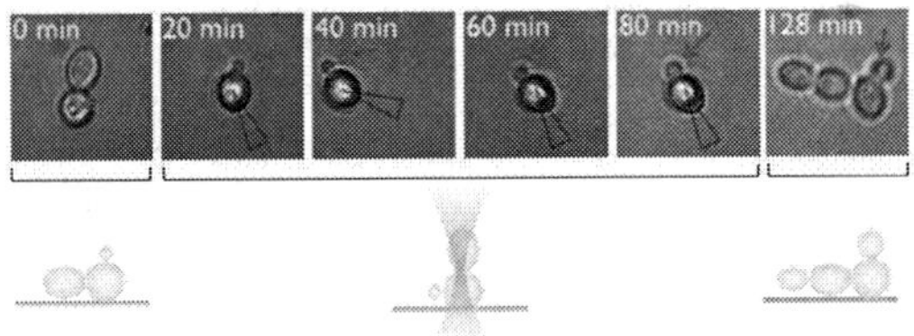

FIGURE 2. Time-lapse images of optically trapped single S. cerevisiae cell using CW Ti:S laser. Black triangular arrow indicates the laser focus. The other arrow indicates the same part of budding.

Next, we observed growth of mixed single yeast cell cultures. Single cultures of *S cerevisiae* and *C albicans* are transferred to a glass chamber with fresh YPD medium. Time-lapse observation of single *S cerevisiae* and *C albicans* is performed as shown in Fig. 3. In this observation, *S cerevisiae* and *C albicans* are not in contact, but both cells can be viewed within the field of view of the microscope. After 80min budding growth is observed in cell both species.

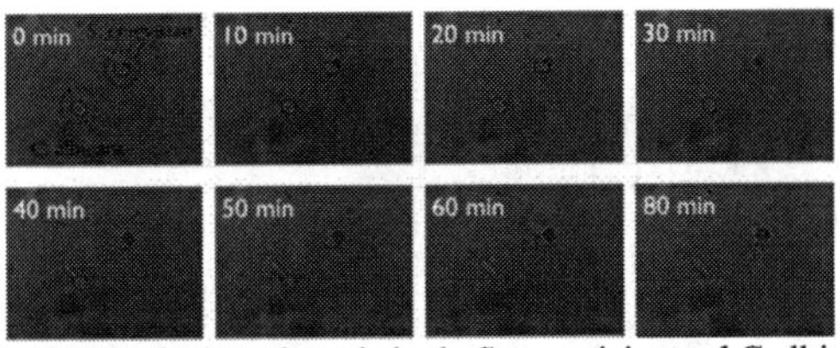

FIGURE 3. Time-lapse images of mix-cultured single S. cerevisiae and C albicans without laser trapping.

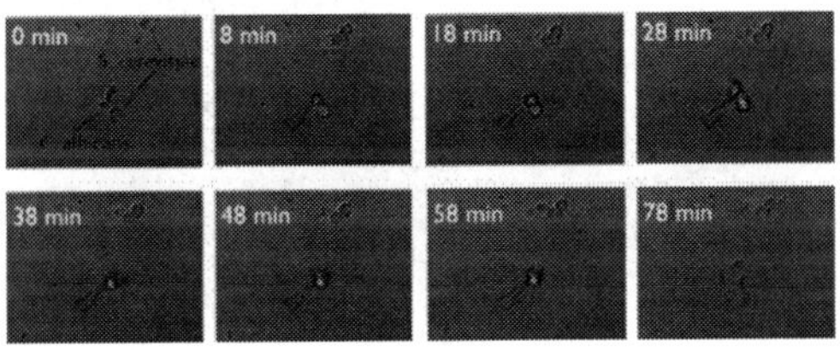

FIGURE 4. Time-lapse images of mixed-cultured single S. cerevisiae cell and C. albicans with CW Ti:S laser trapping. Black arrow indicates the laser focus. Two kinds of yeast cells are forced to be in contact with each other.

Finally, we applied optical trapping to observe the growth of the two cell species that are in contact with each other. Selected single *S cerevisiae* and *C albicans* are simultaneously optically trapped by focused CW mode Ti:S laser at 10 mW of laser power measured at the laser focus. Time-lapse observation is performed as shown in Fig. 4. We note that even after 78 min, budding growth is not observed in both of the cells. As control, we include in the field of view a *S cerevisiae* that is not confined in an optical trap as indicated by the arrow. The unconfined *S cerevisiae* shows budding growth. This result indicates cell-to-cell contact mediated cell growth arrest for two different cell species. This experiment confirms previous reports on contact mediated cell growth arrest using macro-scale analysis[3,4].

CONCLUSIONS

We have observed microscopic analysis or single-cell-level growth of *Saccharomyces cerevisiae* in a mixed cell culture using optical trapping with a near-infrared laser. We confirmed that focused CW mode Ti: Sapphire laser at 10 mW of laser power does not inhibit the yeast cell growth. We then compared cell growth of mixed yeast cell cultures with and without laser trapping. Only when the two different species of yeast cells are in contact with each other, cell growth is inhibited for both species. This result confirms cell-cell interaction mediated cell growth inhibition mechanism.

ACKNOWLEDGMENTS

The project is partially funded by the Philippine Council for Advanced Science and Technology Research and Development (PCASTRD) and the University of the Philippines System grant. The authors also thank W Rivera for the advice about yeast cell culture, and APL Hilario for the technical advice and support on the image acquisition and control software. J Ando thanks the JASSO short-term student exchange promotion program scholarship.

REFERENCES

1. G. M. Heard and G. H. Fleet, *Appl. Environ. Microbiol* **53**, 2171-2174 (1987)
2. K. Machida, T. Tanaka, Y. Yano, S. Otani and M. Taniguchi, *Microbiology* **145**, 293-299 (1999)
3. P. Nissen and N. Arneborg, *Arch Microbiol* **180**, 257-263 (2003)
4. P. Nissen, D. Nielsen and N. Arneorg, *Yeast* **20**, 331 (2003)
5. N. Arneborg and H. Siegumfeldt, G. H. Andersen, P. Nissen, V. R. Daria, P. J. Rodrigo and J. Gluckstad, *FEMS Microbiology Letters* **245**, 155-159 (2005)
6. A. Ashkin, J. M. Dziedzic and T. Yamane, *Nature* **330**, 769-771 (1987)
7. A. Ashkin and J. M. Dziedzic, *Proc. Natl. Acad. Sci. USA* **86**, 7914-7918 (1989)
8. S. M. Block, L. S. B. Goldstein and B. J. Schnapp, *Nature* **348**, 348-352 (1990)
9. M. Zahn and S. Seeger, *Cell Mol. Biol.* **44**, 747-761 (1998)
10. T. N. Buican, M. J. Smyth, H. A. Crissman, G. C. Saltzmann, C. C. Stewart and J. C. Martin, *Applied optics* **26**, 5311-5316 (1987)
11. C. Bustamente, Z. Bryant and S. B. Smith, *Nature* **421**, 423-427 (2003)
12. Yeast cell samples were prepared by the Natural Science Research Center, University of the Philippines, Diliman, Quezon City, Philippines

INTERACTING SYSTEMS

Scaling Features of Violent Social Conflicts

Frederik W. Wiegel

Institute of Theoretical Physics, University of Amsterdam
Valckenierstraat 65, 1018 XE Amsterdam, The Netherlands

Abstract. We discuss two simple quantitative models for: (1) Richardson's law for the distribution of the size of wars; (2) The distribution of the size of severe terrorist attacks. In both cases the models predict power law exponents that are consistent with the empirical data.

Keywords: Scaling law, Richardson's law, Self-organized criticality

PACS: 89.75.Da, 87.23.Ge

POWER LAW DISTRIBUTIONS

Scaling laws are pervasive to the biological world of plants and animals (cf. this author's reviews [1, 2] in the Proceedings of the 3rd and 4th Jagna International Workshops). In recent years it is rapidly becoming clear that many features of social systems are also characterized by power law distributions. This holds especially for some of the most extreme social phenomena: wars and terrorist attacks.

In the following pages we shall discuss two empirical scaling laws: (1) Richardson's law for the distribution of the "size" of a war, and (2) the scaling law for the probability that a terrorist attack will claim k victims. We shall suggest (in the next two sections) two models which enable a quantitative analysis of these phenomena and which predict exponents which are in good agreement with the observed ones.

Before entering into the more mathematical part of the paper, we should like to discuss three issues that are relevant to this type of modelling: (a) The use of "borrowed" nomenclature, like: power law, scaling law, self-organized criticality. (b) The motivation of this author to be also interested in the scaling properties of social systems (instead of biological physiological systems). (c) Recent work of I. Piepers on quantitative (non-scaling) aspects of the International System which shows remarkably regular behavior.

(a) **Nomenclature.** Some of the terminology in the literature has been borrowed from theoretical physics, which sometimes leads to some confusion. If a quantity can take positive real values k = 1, 2, 3,... . and if the probability to find the value k equals

$$p_k = \mathcal{N}^{-1} k^{-\alpha} \qquad (k = 1, 2, 3, ...) \tag{1a}$$

CP1021, *Stochastic and Quantum Dynamics of Biomolecular Systems,*
Proceedings of the 5th Jagna International Workshop, edited by C. C. Bernido and M. V. Carpio-Bernido

with $\alpha>1$ and

$$\mathcal{N} = \sum_{1}^{\infty} k^{-\alpha} \tag{1b}$$

then we shall say that this quantity has a **power law** probability distribution.

Now suppose we somehow change the unit of measurement for this quantity, from unity to λ. The possible values are now λ, 2λ, 3λ,... and the probability $P_{k'}$ to find the values $k' = k\lambda$ equals

$$P_{k'} = p_k = \mathcal{N}^{-1}\lambda^{\alpha}(k')^{-\alpha} = (\mathcal{N}')^{-1}(k')^{-\alpha} \tag{2a}$$

But this is of the same form as (1), apart from a trivial change of the normalization constant to

$$\mathcal{N}' = \mathcal{N}\lambda^{-\alpha} \quad , \tag{2b}$$

which takes care of the new spacing of the possible values. For this reason we shall say that this quantity obeys a **scaling law**: any change of scale leads to the same distribution. In other words: the system itself is not charaterized by an intrinsic scale; hence at any scale the quantity under consideration shows essentially the same behavior.

In physics such behavior is characteristic for a many particle system which is very near to its critical point. This is the reason why the occurence of a power law (1) is often interpreted as a sign of **self-organized criticality**: it is **as if** the system is similar to a thermodynamic system close to its critical point.

(b) **Motivation.** Why should one search for scaling features of social systems, including the violent conflicts that occasionally develop in them? Reference [2] suggests an answer to this question: a souvereign nation can be compared to an organism, and any internal conflict can be compared to a disease. The organism will fight this disease with its immune system; hence one expects similarities between social conflicts and immune reactions, including some similarity in the relevant scaling laws.

(c) **Stability of the International System.** Recently (2006) the dynamics of the International System has been studied by I. Piepers. Although his Ph.D. dissertation has not yet been published, several related papers can be found on the Internet [3]. It is shown here that, for example, the wars fought in Europe during the historical period 1495-1975 show quantitative features of a remarkably regular type. This topic seems to deserve much further analysis.

RICHARDSON'S LAW

Soon after the end of the second World War Richardson studied the way in which the number (S) of battlefield casualties are distributed for a large number of wars for which these data are known. He found that the probability that $S > s$ has the approximate form

$$P(S > s) \simeq Cs^{1-\alpha} \tag{3}$$

where $C > 0$ and $\alpha > 1$. This implies for the probability density $p(s)\ ds$ that a war will have a "size" between s and $s + ds$:

$$p(s) = -\frac{dP}{ds} = C(\alpha - 1)s^{-\alpha}. \tag{4}$$

For a set of wars between 1820 and 1997 the empirical values are

$$\ln C \cong 1.27 \tag{5a}$$

$$\alpha \cong 1.41 \tag{5b}$$

For an even larger set of wars between 1495 and 1965 one finds

$$\alpha \cong 1.57; \tag{5c}$$

c.f. a recent review by Cederman [4]. The empirical scaling law (4) is known as Richardson's law.

In order to model this phenomenon in a simple way we make four assumptions:

(i) At any time during an ongoing war there exists a variable $v \geqslant 0$ which is a quantitative measure for the "willingness to use force."

(ii) When the war starts this quantity has some value $v_0 > 0$.

(iii) The war consists of series of "episodes," each of which claims n_0 battlefield casualties and changes the value of v at random by a small amount $\pm\delta v$.

(iv) The war is over when $v = 0$ for the first time.

One has to calculate the probability that $v = 0$ for the first time after N battles; this equals the probability that a war will have Nn_0 casualties. Obviously, the variable v will perform a free random walk along the continuous v axis, with an absorbing boundary condition at $v = 0$. Denote the diffusion coefficient by D and time by t. The solution to the diffusion equation with the present initial-and boundary conditions can be found in any textbook on diffusion theory:

$$p(v,t) = (4\pi Dt)^{-\frac{1}{2}} \left[\exp\left\{-\frac{(v - v_0)^2}{4Dt}\right\} - \exp\left\{-\frac{(v + v_0)^2}{4Dt}\right\}\right] \tag{6}$$

The probability $P(t)$ that the war is still going at time t is

$$P(t) = \int_0^\infty p\,(v,t)\ dv. \tag{7a}$$

For $t >> v_0^2/D$ the leading term is found to be given by

$$P(t) \cong \frac{v_0}{\sqrt{\pi Dt}}. \tag{7b}$$

Hence the probability density $p(t)\,dt$ that a war will end during the time interval $(t, t+dt)$ is given by

$$p(t) \cong -\frac{dP}{dt} \cong \frac{v_0}{\sqrt{4\pi D}} t^{-\frac{3}{2}} \tag{8}$$

for long times. But this implies that the probability density for a war to have a number of battlefield casualties near to s equals (constant) $s^{-\frac{3}{2}}$. This is Richardson's law (4) with an exponent $\alpha = \frac{3}{2}$.

SCALING LAW FOR TERRORIST ATTACKS

The statistical features of terrorist attacks are the subject of a recent study by Clauset, Young and Gleditsch [5]. They showed that the empirical probability that a terrorist attack claims k victims can be fitted to a power law

$$p(k) \cong (\text{constant}) k^{\alpha} \tag{9}$$

where the exponent α could be estimated separately for the cases of injured victims and deaths:

$$\alpha(\text{injuries}) \cong 2.46 \tag{10a}$$

$$\alpha(\text{deaths}) \cong 2.38 \tag{10b}$$

$$\alpha(\text{total}) \cong 2.48 \tag{10c}$$

One wonders if there might be a simple explanation for this power law.

In this case we make five assumptions, calculate the resulting distribution of number of victims and compare with the empirical data (10). Here are the assumptions:

(i) There exists a "pool" of N single individuals that are "inclined" towards terrorism. We assume N to be large as compared to unity and to be constant in time. (Actually, these individuals are single males, with an age between 17 and 37. The reader might want to consult a paper by Pape [6] for a review of the psychology and strategy that drives suicide terrorism.)

(ii) These individuals form groups of size 2,3,4... .

(iii) Groups grow beacuse two groups can coalesce.

(iv) Any group can disintegrate "spontaneously" into single individuals. Note that we treat N as a constant; this implies that those terrorists that are eliminated as a result of their attacks are instantaneously replaced by an equal number of terrorism-inclined individuals.

(v) At any time, any group can launch an attack, with a probability that is independent of the size or "age" of the group. The number of victims claimed by an attack is proportional to the size of the group that attacks.

In order to derive an equation for the number n_k = 0,1,2,... of groups consisting of $k = 1, 2, 3, \ldots$. individuals we assume that each pair consisting of a group of k and a group of ℓ has a probability $A_0 k^a\, \ell^a$ to coagulate into a single group of size $(k+\ell)$, per unit of time. Here A_0 is a positive constant, and one expects the exponent $a \geqslant 0$ to be near to unity

$$a \cong 1 \tag{11}$$

because the number of human relations between a member of the first group and a member of the second group, equals $k\ell$.

The number of pairs consisting of a k - group and an ℓ - group equals $n_k n_\ell$ for $k = \ell$. It equals $\frac{1}{2}n_k(n_k-1)$ in the case $k=\ell$.If $n_k >> 1$ one can approximate thi~~s~~ by $\frac{1}{2}n_k^2$, which we shall do as it simplifies the mathematics considerably.

The probability, per unit of time, that a k-group disintegrates spontaneously into single individuals will be written $b(k)$. It will turn out that the explicit form of this function is not needed to calculate the equilibrium distribution of the group size, provided N is large as compared to unity.

With these assumptions the equation for the change of $n_k(t)$ with time is given (for $k = 2,3,4,...$) by

$$\begin{aligned}\frac{dn_k}{dt} &= \frac{1}{2}A_0 \sum_{i,j=1}^{\infty}{}' i^a j^a n_i n_j \\ &\quad - A_0 k^a n_k \sum_{j=1}^{\infty} j^a n_j - b(k) n_k,\end{aligned} \tag{12}$$

where the Σ' is over all natural numbers i, j such that $i+j=k$. For the steady state the values of $\lim_{t\longrightarrow\infty} n_k(t)$ will be denoted by n_k^{eq}; they can be solved from the quadratic equations

$$\frac{1}{2}A_0 \sum_{i,j}^{\infty}{}' i^a j^a n_i^{eq} n_j^{eq} = A_0 k^a n_k^{eq} \sum_j j^a n_j^{eq} + b(k) n_k^{eq} \; ; \tag{13}$$

$$(k = 2,3,4,...)$$

The simplest way to solve this set of equations is to introduce the generating functions

$$f(z) \equiv \sum_1^{\infty} k^a n_k^{eq} z^k, \tag{14a}$$

$$g(z) \equiv \sum_1^{\infty} b(k) n_k^{eq} z^k. \tag{14b}$$

Multiply (13) with z^k and sum over k from 2 to ∞ . This gives

$$\frac{1}{2}A_0 f(z) f(z) = A_0 f(1)\{f(z) - n_1^{eq} z\} + g(z), \tag{15}$$

where we used the fact that $b(1) = 0$ because a group of one individual cannot disintegrate into single individuals any further.

The solution of (15) in the general case is a tricky problem, but it is much simpler in our case where z is kept fixed and $N >> 1$. In that case n_k^{eq} will be proportional to N for k smaller than some cut-off k_0 which we need not calculate explicity. Hence the leading order of magnitude of the various terms in the last equation is as follows:

$$f(z) \sim N, \tag{16a}$$

$$g(z) \sim N, \tag{16b}$$

$$\frac{1}{2}A_0 f(z) f(z) \sim N^2, \tag{16c}$$

$$A_0 f(1)\{f(z) - n_1^{eq} z\} \sim N^2. \tag{16d}$$

Hence in this limit (15) can be replaced by

$$\frac{1}{2}f^2(z) - f(1)f(z) + f(1)n_1^{eq} z = 0 \tag{17}$$

which has the solution

$$f(z) = f(1) - \sqrt{f^2(1) - 2f(1)\; n_1^{eq} z}. \tag{18}$$

Substitution of $z = 1$ shows that $f(1) = 2n_1^{eq}$, hence another way to write the generating function is:

$$f(z) = 2n_1^{eq}\{1 - \sqrt{1-z}\}. \tag{19}$$

A glance at the definition (14a) shows that $k^a n_k^{eq}$ now equals the coefficient of z^k in the power series expansion of this function.

For k of order unity these coefficients can easily be calculated "by hand." For $k >> 1$ one can start from Cauchy's theorem, which gives

$$k^a n_k^{eq} = i\frac{n_1^{eq}}{\pi}\oint_C \sqrt{1-z}\frac{dz}{z^{k+1}}, \tag{20}$$

where the contour C encircles the origin of the complex z - plane once in counter-clockwise direction. The contour C can be deformed into a contour C' which encircles the branchcut $1 < z < \infty$ once in a clockwise direction. Write

$$z = 1 + \zeta \tag{21a}$$

and

$$z^{-k-1} \cong \exp\{-(k+1)\zeta\} \tag{21b}$$

for z near to the branchpoint $z = 1$. For a small, positive imaginary part of ζ one has $\sqrt{-\zeta} = -i\sqrt{\zeta}$ and for a small negative imaginary part of ζ one has $\sqrt{-\zeta} = +i\sqrt{\zeta}$. Hence, for $k >> 1$ you find

$$\begin{aligned} k^a n_k^{eq} &\cong \frac{2}{\pi}n_1^{eq}\int_0^\infty \sqrt{\zeta}\exp\{-(k+1)\zeta\}d\zeta \\ &= \frac{1}{\sqrt{\pi}}n_1^{eq}(k+1)^{-\frac{3}{2}}. \end{aligned} \tag{22}$$

The number of groups consisting of k individuals is found in this way:

$$n_k^{eq} \cong \frac{1}{\sqrt{\pi}} n_1^{eq} k^{-a-\frac{3}{2}}, \qquad (k >> 1). \tag{23}$$

Our assumption (v) then implies that the probability that a terrorist attack will claim k victims has a power law distribution of the form (9) with

$$\alpha = a + \frac{3}{2} \tag{24}$$

As a is expected to be near to unity (cf.eq. (11)) the present model predicts the value

$$\alpha \cong \frac{5}{2} \tag{25}$$

which is consisted with the empirical data (10) of Clauset, Young and Gleditsch [5].

CONCLUDING REMARKS

The two preceding paragraphs show that the power law distributions of the size of wars and the severity of terrorist attacks can be well explained by simple stochastic models. This, of course, does not imply that these models are correct; it does increase the motivation to derive further analytical consequences and to compare them with more detailed empirical observations.

Each of the two models has its own peculiar features. For the model of section 2 this feature is expressed by assumption (i) which says that the "willingness to use force" in a conflict can be quantitatively expressed by a positive real number (v). We did not specify how one could actually measure the value of v, and it is not clear if one could measure it at all. This state of affairs (a theory formulated in terms of a mathematical quantity that cannot be measured) would have been unacceptable in physics till 1925, but is now normal for quantum physics (a theory formulated in terms of a complex-valued "probability amplitude" which cannot be measured).

The model of section 3 has the remarkable feature that the exponent α of the power law distribution depends only on the exponent a that determines the coagulation of groups, but is **independent** of the function $b(k)$ that determines the disintegration of existing groups. The system somehow organizes itself in such a way that it generates the same power law distribution, regardless of the details that determine $b(k)$. (Note that one expects $b(k)$ to have considerable effect on the **speed** with which each individual cycles through the clusters of various size). The fact that the power law distribution is independent of $b(k)$ can be viewed as a form of **universality** of the self-organized criticality that is shown by this model.

ACKNOWLEDGEMENTS

The author is indebted to Ingo Piepers and Aaron Clauset for a continuing series of discussions about the quantitative modelling of severe social conflicts, and to the Santa Fe Institute for its frequent hospitality. Many thanks are also due to Christopher Bernido and Victoria Carpio-Bernido for stimulating the scientific interactions between the Philippines and foreign countries, by organizing this ongoing series of International Workshops.

REFERENCES

1. F.W. Wiegel, "Scaling in Biology".In: Functional Integrals in Stochastic and Quantum Dynamics. Eds. C.C. Bernido, M.V. Carpio-Bernido and L. Streit (Central Visayan Institute Foundation, Jagna, 2002 pgs. 32-46.

2. F.W. Wiegel, "Scaling Properties of the Mammalian Immune System". In: Methods and Applications of Infinite Dimensional Analysis. Eds. C.C. Bernido and M.V. Carpio-Bernido (Central Visayan Institute Foundation, Jagna, 2006) pgs. 98 - 113.

3. I. Piepers. Several papers have been placed in the arXiv library of Cornell University (www.arxiv.com; then search under the name "Ingo Piepers"). The most recent one "Self-Organized Characteristics of the International System" can be downloaded freely (use the link http:// arxiv.org/abs/0707.0348). Piepers can also be reached directly by e-mail (ingopiepers@gmail.com).

4. L.E. Cederman "Modelling the Size of Wars". American Political Science Review **97**, 135-150 (2003).

5. A. Clauset, M. Young and K.S. Gleditsch "On the Frequency of Severe Terrorist Events". Journal of Conflict Resolution **51**, 58-87 (2007).

6. R.A. Pape "The Strategic Logic of Suicide Terrorism". American Political Science Review **97**, 343-361 (2003).

Searching for Communities in Bipartite Networks

Michael J. Barber*, Margarida Faria†, Ludwig Streit† and Oleg Strogan†

*Austrian Research Centers GmbH—ARC, Bereich systems research, Vienna, Austria
†CCM, Universidade da Madeira, Funchal, Portugal

Abstract. Bipartite networks are a useful tool for representing and investigating interaction networks. We consider methods for identifying communities in bipartite networks. Intuitive notions of network community groups are made explicit using Newman's modularity measure. A specialized version of the modularity, adapted to be appropriate for bipartite networks, is presented; a corresponding algorithm is described for identifying community groups through maximizing this measure. The algorithm is applied to networks derived from the EU Framework Programs on Research and Technological Development. Community groups identified are compared using information-theoretic methods.

Keywords: Bipartite Networks, Community Structure, R&D Networks, Topical Differentiation
PACS: 89.75.Hc, 89.75.Fb

INTRODUCTION

Networks have attracted a burst of attention in the last decade (useful reviews include Refs. [1–4]), with applications to natural, social, and technological networks. Within biology, networks are prevalent, including: neural networks, where synapses link neurons; metabolic networks, describing metabolic processes in the cell, linking chemical reactions and the regulatory processes that control them; protein interaction networks, representing physical interactions between an organism's proteins; transcription networks, describing regulatory interactions between different genes; food webs, using links to characterize who eats whom; and networks of sexual relations and infections, including AIDS models. Taking a broader view, networks seem to be everywhere! There are: electrical power grids, whose stability relates to the network structure; airline networks, with service efficiency tied to properties of the network; the World Wide Web, with search engines using the network links to locate pages; networks in linguistics, with words linked by co-occurrence; social networks of all sorts [5–7]; collaboration networks, describing joint works amongst actors, authors, research labs, . . . ; and many more.

Of great current interest is the identification of community groups, or modules, within networks. Stated informally, a community group is a portion of the network whose members are more tightly linked to one another than to other members of the network. A variety of approaches [8–16] have been taken to explore this concept; see Refs. [17, 18] for useful reviews. Detecting community groups allows quantitative investigation of relevant subnetworks. Properties of the subnetworks may differ from the aggregate properties of the network as a whole, e.g., modules in the World Wide Web are sets of topically related web pages.

CP1021, *Stochastic and Quantum Dynamics of Biomolecular Systems,*
Proceedings of the 5th Jagna International Workshop, edited by C. C. Bernido and M. V. Carpio-Bernido

Methods for identifying community groups can be specialized to distinct classes of networks, such as bipartite networks [19, 20]. The nodes in a bipartite network can be partitioned into two disjoint sets such that no two nodes within the same set are adjacent. Bipartite networks thus feature two distinct types of nodes, providing a natural representation for many affiliation or interaction networks, with one type of node representing actors and the other representing relations. Examples of actor-relation pairs include people attending events [21–23], court justices making decisions [23], scientists jointly publishing articles [24, 25], organizations collaborating in projects [26, 27], and legislators serving on committees [28]. Arguably, bipartite networks are the empirically standard case for social networks and other interaction networks, with unipartite networks appearing—often implicitly—as projections.

COMMUNITY STRUCTURE IN NETWORKS

A Few Words on Graphs

We formally describe networks using the language of graph theory. Let V be a set of *vertices* and E be a set of vertex pairs or *edges* from $V \times V$. The pair $G = (V, E)$ is called a *graph*. In a *simple graph*, all pairs $\{u, v\} \in E$ are distinct and $\{u, u\} \notin E$, i.e., there are no double lines or loops. Given a partition

$$V = V_1 + V_2 \tag{1}$$

where no edges exist between pairs of points within V_i, $i = 1$ or 2, then G is said to be *bipartite*.

We shall consider simple graphs on a (large) finite set V:

$$V = \{1, 2, \ldots, n\} \quad . \tag{2}$$

The number of edges or *degree* d_v of vertex v is defined by

$$d_v = \#(\{v, \cdot\} \in E) \tag{3}$$

The number of edges is

$$|E| = \frac{1}{2} \sum_{k=1}^{n} d_k \quad . \tag{4}$$

$2|E|$ is often called the *volume* of the graph G.

Graph structure is encoded in the *adjacency matrix*

$$\mathbf{A} = (A_{ik})_{1 \le i,k \le n} \quad , \tag{5}$$

with

$$A_{ik} = \begin{cases} 1 & \text{if } \{i, k\} \in E \\ 0 & \text{otherwise} \end{cases} \quad . \tag{6}$$

Modularity

Detecting community groups allows for the identification and quantitative investigation of relevant subnetworks. Local properties of the community groups may differ from the global properties of the complete network. For example, topically related web pages in the World Wide Web are typically interlinked, so that the contents of pages in distinct community groups should reveal distinct themes. Thus, identification of community groups within a network is a first step towards understanding the heterogeneous substructures of the network.

To identify communities, we take as our starting point the *modularity*, introduced by Newman and Girvan [14]. Modularity makes intuitive notions of community groups precise by comparing network edges to those of a null model. As noted by Newman [29]:

> A good division of a network into communities is not merely one in which there are few edges between communities; it is one in which there are fewer than expected edges between communities.

Definition 1 *The modularity Q is—up to a normalization constant—the number of edges within communities c minus those for a null model:*

$$Q \equiv \frac{1}{2|E|} \sum_c \sum_{i,j \in c} \left(A_{ij} - P_{ij} \right) \quad . \tag{7}$$

Along with eq. (7), it is necessary to provide a null model, defining P_{ij}. The standard choice for the null model constrains the degree distribution for the vertices to match the degree distribution in the actual network. Random graph models of this sort are obtained [30] by putting an edge between vertices i and j at random, with the constraint that on average the degree of any vertex i is d_i. This constrains the expected adjacency matrix such that

$$d_i = E\left(\sum_j A_{ij} \right) \quad . \tag{8}$$

Denote $E\left(A_{ij}\right)$ by P_{ij} and assume further that P factorizes into

$$P_{ij} = p_i p_j \quad , \tag{9}$$

leading to

$$P_{ij} \equiv \frac{d_i d_j}{2|E|} \quad . \tag{10}$$

A consequence of the null model choice is that $Q = 0$ when all vertices are in the same community.

The goal now is to find a division of the vertices into communities such that the modularity Q is maximal. An exhaustive search for a decomposition is out of the question: even for moderately large graphs there are far too many ways to decompose them into communities. Fast approximate algorithms do exist (see, for example, Refs. [31, 32]).

Finding Communities with BRIM

Specific classes of networks have additional constraints that can be reflected in the null model. For bipartite graphs, the null model should be modified to reproduce the characteristic form of bipartite adjacency matrices

$$\mathbf{A} = \begin{bmatrix} \mathbf{O} & \mathbf{M} \\ \mathbf{M}^{\mathrm{T}} & \mathbf{O} \end{bmatrix} . \tag{11}$$

Recently, specialized modularity measures and search algorithms have been proposed for finding communities in bipartite networks [19, 20].

We make use of the algorithm called BRIM: bipartite, recursively induced modules [19]. Starting from a (more or less) ad hoc partition of the vertices of type 1, it is straightforward to optimize a corresponding decomposition of the vertices of type 2. From there, optimize the decomposition of vertices of type 1, and iterate. In this fashion, modularity increases until a (local) maximum is reached. However, the question remains: is the maximum a "good" one? At this level then a random search is called for, varying the composition and number of communities, with the goal of reaching a better maximum after a new round of hill climbing using the BRIM algorithm.

AN INITIAL LOOK AT EU RESEARCH NETWORKS

In the ongoing research project NEMO [33], networks of research and development collaborations under EU Framework Programs FP1, FP2,..., FP5 are studied. The collaborations in the Framework Programs give rise to bipartite graphs, with edges existing between projects and the organizations which take part in them. With this construction, participating organizations are linked only through joint projects.

In the various Framework Programs, the number of organizations ranges from 2,000 to 20,000, the number of projects ranges from 3,000 to 15,000, and the number of links between them (project participations) ranges from 10,000 to 250,000 (see table 1 for precise values). A popular approach in social network analysis—where networks are often small, consisting of a few dozen nodes—is to visualize the networks and identify community groups by eye. However, the Framework Program networks are much larger: can we "see" the community groups in these networks?

Structural differences or similarities of such networks are not obvious at a glance. For a graphical representation of the organizations and/or projects by dots on an A4 sheet of paper, we would have to put these dots at a distance of about 1 mm from each other, and we then still would not have drawn the links (collaborations) which connect them.

Previous studies used a list of coarse graining recipes to compact the networks into a form which would lend itself to a graphical representation [34]. As an alternative we have attempted to detect communities just using BRIM, i.e., purely on the basis of relational network structure, and blind with respect to any additional information about the nature of agents.

In Fig. 1, we show a community structure for FP3 found using the BRIM algorithm, with a modularity of $Q = 0.602$ for 14 community groups. The communities are shown as vertices in a network, with the vertex positions determined using spectral methods

TABLE 1. R&D projects and participants in the EU Framework Programs. The projects and organizations are used to define bipartite networks for each of the Framework Programs.

Framework Program	Period	Projects	Organizations
FP1	1984–1987	3,283	1,981
FP2	1987–1981	3,885	4,572
FP3	1990–1994	5,529	7,324
FP4	1994–1998	15,061	19,755
FP5	1998–2002	15,559	22,303

[35]. The area of each vertex is proportional to the number of edges from the original network within the corresponding community. The width of each edge in the community network is proportional to the number of edges in the original network connecting community members from the two linked groups. The vertices and edges are shaded to provide additional information about their topical structure, as described in the next section. Each community is labeled with the most frequently occurring subject index.

Topical Profiles of Communities

Projects are assigned one or more standardized subject indices. There are 49 subject indices in total, ranging from *Aerospace* to *Waste Management*.

We denote by

$$f(t) > 0 \tag{12}$$

the frequency of occurrence of the subject index t in the network, with

$$\sum_t f(t) = 1 \quad . \tag{13}$$

Similarly we consider the projects within one community c and the frequency

$$f_c(t) \geq 0 \tag{14}$$

of any subject index t appearing in the projects only of that community. We call f_c the topical profile of community c to be compared with that of the network as a whole.

Topical differentiation of communities can be measured by comparing their profiles, among each other or with respect to the overall network. This can be done in a variety of ways [36], such as by the Kullback “distance”

$$D_c = \sum_t f_c(t) \ln \frac{f_c(t)}{f(t)} \quad . \tag{15}$$

A true metric is given by

$$d_c = \sum_t |f_c(t) - f(t)| \quad , \tag{16}$$

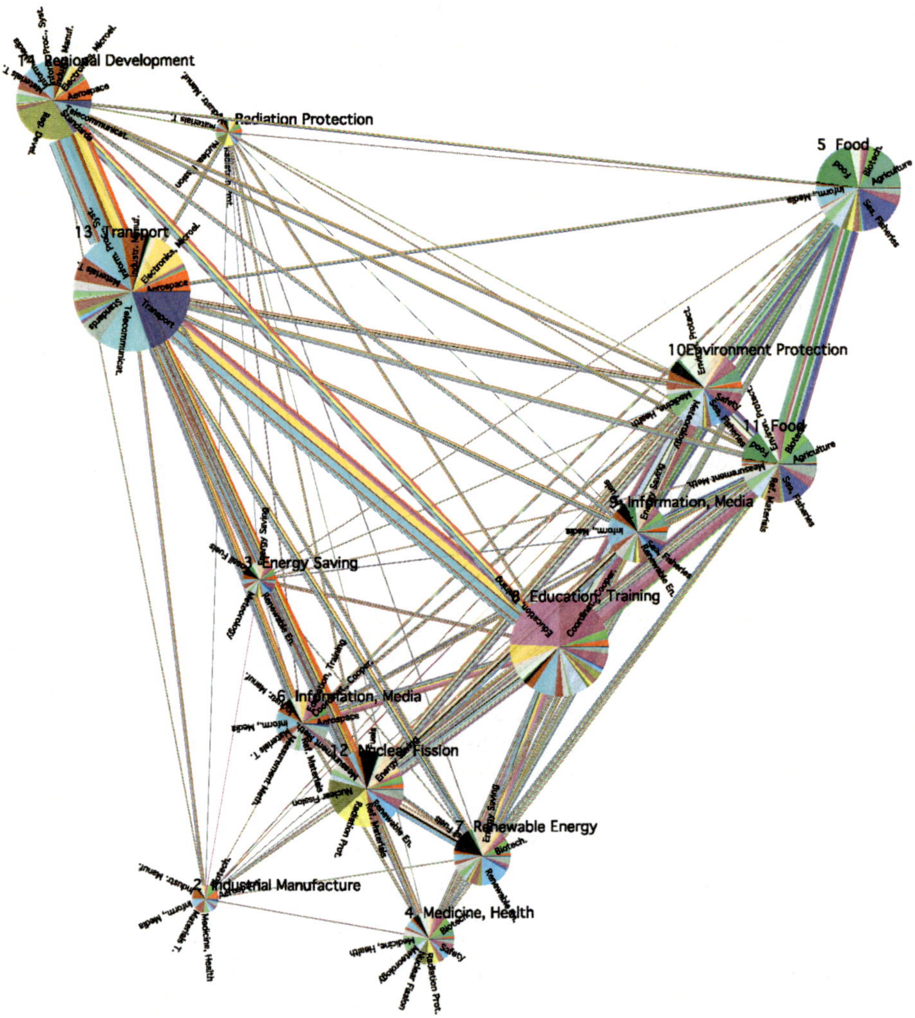

FIGURE 1. Community groups in the network of projects and organizations for FP3.

ranging from zero to two.

Topical differentiation is illustrated in Fig. 2. In the figure, example profiles are shown, taken from the network in Fig. 1. The community-specific profile corresponds to the community labeled '11. Food" in Fig. 1. Based on the most frequently occurring subject indices—*Agriculture, Food,* and *Resources of the Seas, Fisheries*—the community consists of projects and organizations focussed on R&D related to food products. The topical differentiation is $d_c = 0.90$ for the community shown.

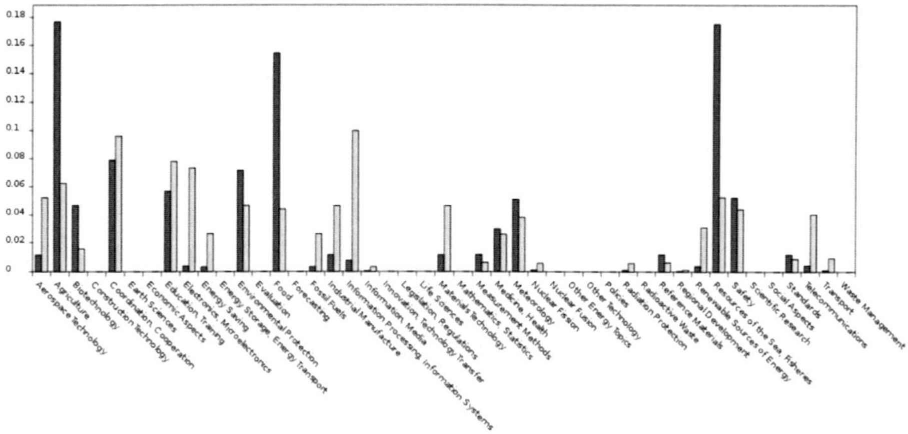

FIGURE 2. Topical differentiation in a network community. The histogram shows the difference between the topical profile $f_c(t)$ for a specific community (dark bars) and the overall profile $f(t)$ for the network as a whole (light bars). The community-specific profile shown is for the community labeled "11. Food" in Fig. 1. The community has $d_c = 0.90$.

Projects and Subject Indices

For further analysis, we have also looked for communities in networks of projects and subject indices. Here, the projects and subject indices constitute the vertices of a bipartite network, with edges existing between projects and the subject indices assigned to them. This construction disregards the organizations, providing an alternate approach to investigating the topical structure of the Framework Programs.

In Fig. 3, we show a set of nine community groups identified for the network of projects and subject indices in FP5. The modularity is $Q = 0.50290$ with these community groups.

The later Framework Programs, such as FP5, show a fair degree of overlap between the communities, due to the subject indices being freely assigned in project applications. This is in marked contrast to the networks for the first three Framework Programs, where topics were attributed rigidly within thematic subprograms. The communities for FP1–3 thus have more clearly segregated community structures. FP1 is particularly extreme, having *no* overlaps between the communities. The differences between Framework Programs point to the need for some care in interpreting community structures: the communities in FP1–3 reflect policy structures while those found for later Framework Programs are more representative of interaction patterns.

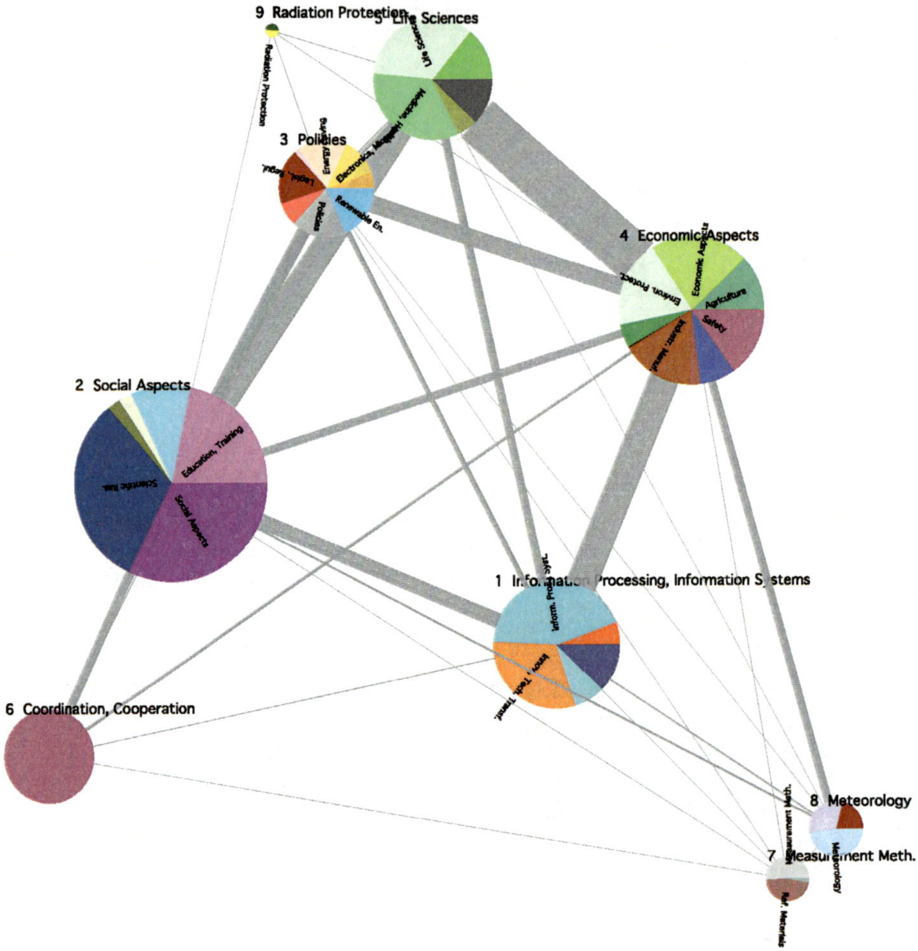

FIGURE 3. Community groups in the network of projects and subject indices for FP5.

Comparing Decompositions with the Mutual Information

For different decompositions of a (finite) set Ω, a measure of similarity is provided by the *normalized mutual information* [17]. Set

$$p(A) = \frac{|A|}{|\Omega|} \quad . \tag{17}$$

For divisions of Ω into

1. J disjoint subsets $X_1, \ldots, X_J$ and
2. K disjoint subsets $Y_1, \ldots, Y_K$,

compute the mutual information

$$I_0(X,Y) = \sum_{jk} p\left(X_j \cap Y_k\right) \log \frac{p\left(X_j \cap Y_k\right)}{p\left(X_j\right) p\left(Y_k\right)} \tag{18}$$

and normalize

$$I(X,Y) = \frac{2I(X,Y)}{H(X)+H(Y)} \quad , \tag{19}$$

where the entropy H is

$$H(X) = -\sum_{j} p\left(X_j\right) \log p\left(X_j\right) \quad . \tag{20}$$

With the definitions in eqs. (18) through (20), I ranges from zero, for uncorrelated decompositions of the set, to one, for perfectly correlated decompositions.

Using the BRIM algorithm, we have partitioned network vertices into community groups by maximizing the modularity Q. In principle, many dissimilar partitions of the vertices could produce similar modularity values. With the normalized mutual information (or a similar measure), we can assess the amount of shared structure between two different partitions of the vertex set.

For example, in Fig. 3 we have shown a decomposition of the network of projects and subject indices for FP5. In Fig. 4, we show a second decomposition of the same network. The modularity is nearly identical—0.50296 instead of the $Q = 0.50290$ seen previously—and the decompositions have some visible similarities. However, there are also definite structural differences, most prominently that the second decomposition has only eight communities while the first has nine. Are the structural differences significant? The normalized mutual information is found to be $I = 0.98$, indicating a strong correlation between the two decompositions and demonstrating that they have relatively minor structural differences.

Outlook

We have successfully identified community groups in networks defined from the Framework Program projects and the subject indices assigned to them. The full networks defined from the projects and the organizations taking part in them are considerably larger and correspondingly more challenging to investigate. For the full organizations-projects network, the BRIM hill climbing algorithm is being supplemented with an aggressive, probabilistic search through community configurations upon which BRIM acts. This extended search has only just begun, but preliminary results are encouraging.

Once established, communities will then be investigated with regard to their internal structure, with the goal of identifying correlated properties within communities and contrasting properties across communities. We expect the analysis of internal structure to reveal patterns, themes, and motivations of collaborative research and development in the European Union.

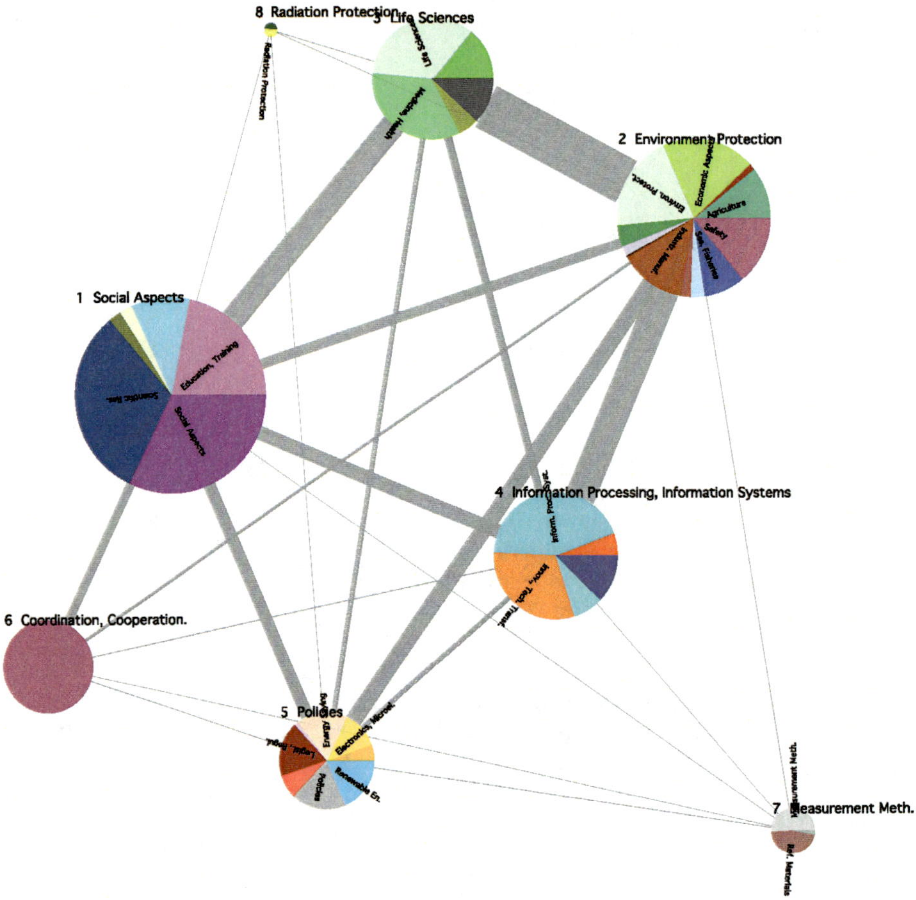

FIGURE 4. An alternate set of community groups in the network of projects and subject indices for FP5.

ACKNOWLEDGMENTS

The authors gratefully acknowledge financial support from the European FP6-NEST-Adventure Programme, under contract number 028875; from the Austrian Research Centers, in the *Leitprojekt "Bewertung und Gestaltung von Wissensnetzwerken"*; from the Portuguese FCT, under projects POCTI/MAT/58321/2004/FSE-FEDER and FCT/POCTI-219/FEDER; and from the Alexander von Humboldt Foundation, for travel support.

REFERENCES

1. C. Christensen, and R. Albert, *International Journal of Bifurcation and Chaos* **17**, 2201–2214 (2007), URL http://arxiv.org/abs/q-bio.OT/0609036, special Issue "Complex Networks' Structure and Dynamics".
2. S. N. Dorogovtsev, and J. F. F. Mendes, "The shortest path to complex networks," in *Complex Systems and Inter-disciplinary Science*, edited by N. Johnson, J. Efstathiou, and F. Reed-Tsochas, World Scientific, 2004, URL http://arxiv.org/abs/cond-mat/0404593.
3. M. E. J. Newman, *SIAM Review* **45**, 167–256 (2003), URL http://arxiv.org/abs/cond-mat/0303516.
4. R. Albert, and A.-L. Barabási, *Reviews of Modern Physics* **74**, 47 (2002), URL http://link.aps.org/abstract/RMP/v74/p47.
5. R. S. Burt, *Structural Holes: The Social Structure of Competition*, Harvard University Press, Cambridge, MA, 1992.
6. F. Coulon, The use of social network analysis in innovation research: A literature review, Danish Research Unit for Industrial Dynamics (2005).
7. M. Granovetter, *The American Journal of Sociology* **78**, 1360–1380 (1973).
8. L. Angelini, S. Boccaletti, D. Marinazzo, M. Pellicoro, and S. Stramaglia, *Chaos: An Interdisciplinary Journal of Nonlinear Science* **17**, 023114 (2007), URL http://arxiv.org/abs/cond-mat/0610182.
9. V. Gol'dshtein, and G. A. Koganov, An indicator for community structure (2006), URL http://arxiv.org/abs/physics/0607159, preprint.
10. M. B. Hastings, *Physical Review E (Statistical, Nonlinear, and Soft Matter Physics)* **74**, 035102 (2006), URL http://arxiv.org/abs/cond-mat/0604429.
11. M. E. J. Newman, and E. A. Leicht, *PNAS* **104**, 9564–9569 (2007), URL http://www.pnas.org/cgi/content/abstract/104/23/9564, http://arxiv.org/abs/physics/0611158/.
12. J. Reichardt, and S. Bornholdt, *Physical Review E (Statistical, Nonlinear, and Soft Matter Physics)* **74**, 016110 (2006), URL http://link.aps.org/abstract/PRE/v74/e016110, cond-mat/0603718.
13. G. Palla, I. Derenyi, I. Farkas, and T. Vicsek, *Nature* **435**, 814–818 (2005), URL http://arxiv.org/abs/physics/0506133.
14. M. E. J. Newman, and M. Girvan, *Physical Review E (Statistical, Nonlinear, and Soft Matter Physics)* **69**, 026113 (2004), URL http://link.aps.org/abstract/PRE/v69/e026113.
15. A. Clauset, M. E. J. Newman, and C. Moore, *Physical Review E (Statistical, Nonlinear, and Soft Matter Physics)* **70**, 066111 (2004), URL http://link.aps.org/abstract/PRE/v70/e066111.
16. M. Girvan, and M. E. J. Newman, *PNAS* **99**, 7821–7826 (2002), URL http://www.pnas.org/cgi/content/abstract/99/12/7821, http://www.pnas.org/cgi/reprint/99/12/7821.pdf.
17. L. Danon, A. Díaz-Guilera, J. Duch, and A. Arenas, *J. Stat. Mech.* **P09008** (2005), URL http://www.iop.org/EJ/article/1742-5468/2005/09/P09008/jstat5%_09_p09008.html.
18. M. E. J. Newman, *Eur. Phys. J. B* **38**, 321–330 (2004), URL http://www-personal.umich.edu/~mejn/papers/epjb.pdf.
19. M. J. Barber, *Physical Review E (Statistical, Nonlinear, and Soft Matter Physics)* **76**, 066102 (2007), URL http://arxiv.org/abs/0707.1616, arXiv:0707.1616.
20. R. Guimerà, M. Sales-Pardo, and L. A. N. Amaral, *Physical Review E (Statistical, Nonlinear, and Soft Matter Physics)* **76**, 036102 (2007), URL http://link.aps.org/abstract/PRE/v76/e036102.
21. A. Davis, B. B. Gardner, and M. R. Gardner, *Deep South*, University of Chicago Press, 1941.
22. L. Freeman, "Finding social groups: A meta-analysis of the southern women data," in *Dynamic Social Network Modeling and Analysis*, edited by R. Breiger, K. Carley, and P. Pattison, The National Academies Press, Washington, DC, 2003, URL http://moreno.ss.uci.edu/85.pdf.
23. P. Doreian, V. Batagelj, and A. Ferligoj, *Social Networks* **26**, 29–53 (2004).
24. M. E. J. Newman, *Physical Review E (Statistical, Nonlinear, and Soft Matter Physics)* **64**, M. E.

J. Newman (2001), URL http://www-personal.umich.edu/~mejn/papers/016131.pdf.
25. M. E. J. Newman, *Physical Review E (Statistical, Nonlinear, and Soft Matter Physics)* **64**, 016132 (2001), URL http://www-personal.umich.edu/~mejn/papers/016132.pdf.
26. M. J. Barber, A. Krueger, T. Krueger, and T. Roediger-Schluga, *Physical Review E (Statistical, Nonlinear, and Soft Matter Physics)* **73**, 036132 (2006), URL http://link.aps.org/abstract/PRE/v73/e036132.
27. T. Roediger-Schluga, and M. J. Barber, *IJFIP* **4**, 321–347 (2008), special Issue on "Innovation Networks".
28. M. A. Porter, P. J. Mucha, M. E. J. Newman, and A. J. Friend, *Physica A* **386**, 414–438 (2007), URL http://arxiv.org/abs/physics/0602033.
29. M. E. J. Newman, *Proceedings of the National Academy of Sciences* **103**, 8577–8582 (2006), URL http://www.pnas.org/cgi/content/abstract/103/23/8577.
30. F. Chung, and L. Lu, *Annals of Combinatorics* **6**, 125–145 (2002).
31. J. M. Pujol, J. Bejar, and J. Delgado, *Physical Review E (Statistical, Nonlinear, and Soft Matter Physics)* **74**, 016107 (2006), URL http://link.aps.org/abstract/PRE/v74/e016107.
32. M. E. J. Newman, *Physical Review E (Statistical, Nonlinear, and Soft Matter Physics)* **69**, 066133 (2004), URL http://link.aps.org/abstract/PRE/v69/e066133.
33. NEMO: Network models, governance and R&D collaboration networks (????), URL http://www.nemo-net.eu.
34. S. Breschi, and L. Cusmano, *International Journal of Technology Management* **27**, 747–72 (2004).
35. A. J. Seary, and W. D. Richards, "Spectral methods for analyzing and visualizing networks: an introduction," in *Dynamic Social Network Modeling and Analysis: Workshop Summary and Papers*, edited by R. Breiger, K. Carley, and P. Pattison, The National Academies Press, Washington, D.C., 2003, pp. 209–228, URL http://www.sfu.ca/~richards/Pages/NAS.AJS-WDR.pdf.
36. A. Gibbs, and F. Su, *International Statistical Review* **70**, 419–435 (2002).

To Ludwig Streit on his 70th birthday with warm wishes

Exploring Urban Environments By Random Walks

Ph. Blanchard and D. Volchenkov

Bielefeld-Bonn Stochastic Research Center (BiBoS),
Bielefeld University, Postfach 100131, 33501 Bielefeld, Germany email

Abstract. A complex web of roads, walkways and public transport systems can hide areas of geographical isolation very difficult to analyze. Random walks are used to spot the structural details of urban fabric.

Keywords: Space syntax, Queuing networks, Random walks
PACS: 89.65.Lm, 89.75.Fb, 05.40.Fb, 02.10.Ox

INTRODUCTION

In the present paper, we discuss the application of Markov processes to urban studies. However, it is important to emphasize that the approach related to Markov processes (such as random walks) can be used to analyze any type of complex networks. The main technical idea beyond the method is to investigate the properties of networks by exploiting the spectral properties of self-adjoint operators (symmetric matrices) defined on their graph representations. It is then obvious that a network can be effectively investigated by spectral methods, if it is not too large. The spatial networks of human settlements give good examples - for instance, the city of Paris (inside the Periphery Boulevard) contains only 5131 interconnected open spaces.

A city converts a space pattern into a pattern of relationships. While discussing the impact a spatial structure has on the social establishment, it is inevitable to mention two Englishmen. After the British House of Commons had been destroyed in 1941, a long debate commenced in the Parliament on whether the old House insufficient to seat for all of its members be rebuilt or a new modern House should be constructed. In his speech to the meeting in the House of Lords, October 28, 1943, Sir Winston Churchill exclaimed, "We shape our buildings, and afterwards our buildings shape us" [1], as if he believed that the movement pattern emerged in the old House of Common is responsible for the establishing and sustain of the British political tradition. This intuitive knowledge had been conceived by another British gentleman, Bill Hillier, professor at University College of London, within the concept of space syntax, a theory developed in the late 1970s, that seeks to reveal the mutual effects of complex spatial urban networks on society and vice versa [2, 3]. The joint use of scarce space by many people creates life in cities which is driven largely by micro-economic factors which are invariant over various cultures that tends to give cities similar structures. Hereby, the street configuration itself naturally creates differential patterns of occupancy (as an emergent phenomenon),

CP1021, *Stochastic and Quantum Dynamics of Biomolecular Systems,*
Proceedings of the 5th Jagna International Workshop, edited by C. C. Bernido and M. V. Carpio-Bernido

whereby some streets become, over time, more highly used than others [4]. At the same time, a background residential space process driven by cultural factors tend to make cities different from each other, so that the emergent urban grid pattern containing an imprint of certain geometrical regularities forms a network of interconnected open spaces, being a historical record of a city creating process driven by human activity and containing traces of society and history [2].

It is important to note that from the very beginning, graph theory had been developed in a close relation to urban studies: Leonard Euler's founding paper on the Seven Bridges of Königsberg published in 1736 is regarded as the first paper in the history of graph theory, [5]. In the present paper, we review the spectral algorithms designed to analyze transport networks. In particular, the method proves its efficiency for revealing the disadvantageous areas designed to foster crime, deprivation, and ghettoisation, [6].

In the forthcoming sections, we explain that city space syntax can be considered as a queuing network operating in the free flow regime (see Sec.). The steady state of a queuing network (see Sec.) is determined by the set of automorphisms of the graph spanning the network (Sec.). We shall show that these automorphisms naturally correspond to random walks of the nearest neighboring type, in which a traveller, after passing through an open space, randomly chooses another space to move in among those intersecting with the given one. How many possibilities to move the walker has while in an arbitrary place? - Indeed, it depends upon the type of the city he dwells in. We have analyzed two German medieval organic cities, the city canal network in Venice, and the regular street grid in Manhattan and found that while the organic towns are relatively compact (the majority of streets in them are just three steps apart from each other), the bigger urban patterns demonstrate the structure of small worlds, in which a dense network of relatively short local streets is supported by the long axial itineraries connecting the far-away districts within the city that essentially simplifies the navigation and transportation tasks in the urban environment (see Sec.). Being a representation of the set of graph automorphisms, random walks provide us with an effective tool for describing the graph structure in details (see Sec. ,). In Sec. , we illustrate the method by studying two graphs by random walks. The first graph is the regular Petersen graph which contains just 10 nodes, and we investigate the network of Venetian 96 canals as the second example.

The term "random walk" was originally proposed by K. Pearson in 1905 in his letter to *Nature* devoted to a simple model describing a mosquito infestation in a forest: at each time step, a single mosquito moves a fixed length, at a randomly chosen angle (see [7]). Nowadays it is well known that random walks could be effectively used in order to investigate and characterize how effectively the nodes and edges of large networks can be covered by different strategies, [8, 9, 10]. Random walks help us to explore the structural properties of complex urban networks which cannot be detected by the usual methods designed primarily to unravel the structures of large quasi-random graphs. We conclude in the last section (Sec.).

CITY SPACE SYNTAX AS A QUEUEING NETWORK WITH A STEADY-STATE BEHAVIOR

Any graph representation naturally arises as the outcome of a categorization, when we abstract a real world system by eliminating all but one of its features and by grouping together items (or places) sharing a common attribute. For instance, the common attribute of all open spaces in city space syntax is that we can move through them. All elements called nodes that fall into one and the same set V are considered as essentially identical; permutations of them within V are of no consequence. The symmetric group $\mathbb{S}_N$ consisting of all permutations of N elements (N being the cardinality of the set V) constitute therefore the symmetry group of V. If we denote by $E \subseteq V \times V$ the set of ordered pairs of nodes called edges, then a graph is a map $G(V,E) : E \to K \subseteq \mathbb{R}_+$. We suppose that the graph is undirected and has no multiple edges. If two nodes are adjacent, $(i,j) \in E$, we write $i \sim j$. The degree of the node $i \in V$ is the number of its neighbors in G, $\deg_G(i) \quad k_i$.

The set of interconnected urban open spaces in which a pedestrian or a vehicle, upon departures from one space, can join another constitutes a queueing network that can be regarded as a graph whose nodes represent the open spaces (streets, squares, and roundabouts), and whose edges represent links between nodes (junctions and crossroads). In the queueing theory [11], the paths along which a traveller may move from service station to service station are determined by the routing probabilities, and then the theory of Markov chains provides the statistical models for the analysis of queueing networks. Travellers arriving to an open space are either moving through it immediately or queuing until the space becomes available. Once the space is passed through, the traveller is routed to its next station, which is chosen according to a probability distribution among all other open spaces linked to the given one in the city. However, if the destination space has finite capacity then it may be full and then the traveller will be blocked at its current location until the next space becomes available.

Steady states in queuing networks

If the network of N open spaces is not at all congested (as long as inflows are compatible with system capacities), there is a network flow - say, cars are travelling through it at some constant speed $v > 0$ in the free flow regime. Given $(X_t : t \geq 0)$, a i.i.d. random process characterized by some distribution f defined on V and let $p_i > 0$ be the arrival probability at the node $i \in V$, in a sequence of n independent Bernoulli trials, then travellers visit the node i in average np_i times. If trials are conducted in a finite time interval t, during which events occur in sequence, we denote the proportion of time spent by a traveller in i by $\lambda_i > 0$,

$$np_i \quad \lambda_i t, \tag{1}$$

The above formula expresses the linear Ohm's law relating the time variable with the number of independent Bernoulli trials, $t \quad (p_i/\lambda_i)n$.

Assuming the Kirchoff's current law, i.e. the total flow into any node equals the flow out, then the proportions λ_i can be obtained as the solutions of the global balance equations (transport equations) satisfying a normalization condition,

$$\sum_{i \in V} \lambda_i \quad 1. \tag{2}$$

In the queuing network load in the stationary regime with free flow, the steady states (or densities) $\lambda_i > 0$ can be interpreted as the length distances of the correspondent streets $i \in V$ normalized by (2).

In relation to a given distribution of λ_i, there are several "performance indicators" characterizing the transport properties of the queuing network with respect to the random process $(X_t : t \geq 0)$. Given the arrival probability $p_i \quad t\lambda_i/n$, no matter from which node $X_{t=0} \quad j$ the process starts, the first arrival probability (FAP) that a traveller arrives at the node $i \in V$ by time $t' > t$ is given by $P(t' > t \,|\, X_{t'} \quad i) \quad (1 - t \cdot \lambda_i/n)^n$. First encounter properties estimated by random walks play a crucial role in explorations of various real-world networks, including epidemic spreading, transport in disordered media, and neuron firing dynamics (see [12] and references therein). In the limit of an infinite number of trials $n \to \infty$, the FAP is asymptotically exponential,

$$\lim_{n \to \infty} P\left(t' > t \,\middle|\, X_{t'} \quad i\right) \quad e^{-\lambda_i t}, \tag{3}$$

provided $(X_t : t \geq 0)$ preserves the proportion of sojourn time λ_i given by (1). An exponential distribution arises naturally when modelling the time between independent events that happen at a constant average rate. The typical recurrence time which characterizes the random duration elapsed between two consequent arrivals of the traveller at $i \in V$ in the free flow regime described by the process $(X_t : t \geq 0)$ is accounted by the probability density function (pdf) $F_i(t) \quad \lambda_i e^{-\lambda_i t}$. Then the expected first-passage time (FPT) to the node $i \in V$ from a node randomly chosen among all nodes in the networks is given by

$$\frac{1}{\lambda_i} \quad \int_0^{\infty} t\, F_i(t)\, dt \tag{4}$$

In complex networks, the value of FPT crucially depends on the confining environment and is not determined by the simple relation (4). The FPT has been recently recognized as a key quantity to specifying the transport-limited kinetics in many models of transport in complex media, [12].

Other characteristic time can be used in order to quantify the long-run properties of the entire network. Let $\mathfrak{T} \quad \min\left(t : \bigcup_{\tau=0}^{t} \{X_\tau\} \quad V\right)$ be the covering time (also: the coupon-collector's time) required the traveller visits all nodes of the closed queuing network (say, rides along all streets in the city). The value of $\mathfrak{T}$ crucially depends upon the local transport properties of the network, [13]. It may be that $\mathfrak{T} \quad \infty$ if the graph spanning the network is directed, and the process $(X_t : t \geq 0)$ defined on that is irreversible. The typical value of $\mathfrak{T}$ for a finite queuing network is given by the unique solution τ of the equation

$$\sum_{i \in V} e^{-\lambda_i \tau} \quad 1, \tag{5}$$

expressing the idea that after the tour the traveller inevitably returns into one of the previously visited nodes [13]. It is easy to demonstrate that the minimal typical value,

$\tau_{\min} \quad N \log N$, is achieved for the uniform distribution $\lambda_i \quad N^{-1}$. It is also clear that $\tau \to \infty$ if the network contains least frequently visited nodes for which $\lambda_i \to 0$.

However, in many situations, the arrival probabilities p_i are not independent. It is intuitively clear that in real transport networks, the arrival rates p_i at the different nodes are correlated in accordance to the chance to be joined by a path in the graph and therefore may be sensitive to the particular graph symmetries. In the forthcoming subsection, we derive the distribution λ_i associated to the set of automorphisms of a simple graph G.

Graph automorphisms and steady states

Among all measures which can be defined on the set of nodes V, the set of normalized measures (densities) $\lambda_i \geq 0$, $i \in V$, satisfying (2), are of essential interest since they express the conservation of a quantity and therefore may be associated to some physical process.

The fundamental physical process defined on the graph $G(V,E)$ is generated by the subset of its automorphisms preserving the degrees of all nodes, k_i.

For each graph $G(V,E)$, there exists a unique, up to permutations of rows and columns, adjacency matrix $\mathbf{A}$ identified with a linear endomorphism of the vector space of all functions from V into $\mathbb{R}$. In the special case of a finite simple graph (an undirected graph that has no self-loops), the adjacency matrix is a $(0,1)$-matrix such that $A_{ij} \quad 1$ if $i \,/\, j$, $i \sim j$, and $A_{ij} \quad 0$ otherwise. The degree of a node $i \in V$ is therefore given by

$$k_i \quad \sum_{i \sim j} A_{ij}. \tag{6}$$

The set of graph automorphisms, the mappings of the graph to itself preserving all of its structure, is specified by the symmetric group $\mathbb{S}_N$ including all admissible permutations $\Pi \in \mathbb{S}_N$ taking the node $i \in V$ to $\Pi(i) \in V$. The representation of $\mathbb{S}_N$ consists of all $N \times N$ matrices Π_Π, such that $(\Pi_\Pi)_{i,\Pi(i)} \quad 1$, and $(\Pi_\Pi)_{i,j} \quad 0$ if $j \,/\, \Pi(i)$. A linear transformation of the adjacency matrix

$$Z(\mathbf{A})_{ij} \quad \sum_{s,l=1}^{N} \mathscr{F}_{ijsl} A_{sl}, \quad \mathscr{F}_{ijsl} \in \mathbb{R} \tag{7}$$

is a graph automorphism if

$$\Pi_\Pi^{\mathrm{T}} Z(\mathbf{A})\, \Pi_\Pi \quad Z\left(\Pi_\Pi^{\mathrm{T}} \mathbf{A} \Pi_\Pi\right), \tag{8}$$

for any $\Pi \in \mathbb{S}_N$. The latter relation is satisfied if the entries of the tensor $\mathscr{F}$ in (7) meets the following symmetry property:

$$\mathscr{F}_{\Pi(i)\,\Pi(j)\,\Pi(s)\,\Pi(l)} \quad \mathscr{F}_{ijsl}, \tag{9}$$

for any $\Pi \in \mathbb{S}_N$. The action of permutations preserves the conjugate classes of index partition structures, so that any appropriate tensor $\mathscr{F}$ satisfying (9) can be expressed as a linear combination of the following tensors: $\{1, \delta_{ij}, \delta_{is}, \delta_{il}, \delta_{js}, \delta_{jl}, \delta_{sl}, \delta_{ij}\delta_{js}, \delta_{js}\delta_{sl}, \delta_{sl}\delta_{li}, \delta_{li}\delta_{ij}, \delta_{ij}\delta_{sl}, \delta_{is}\delta_{jl}, \delta_{il}\delta_{js}, \delta_{ij}\delta_{il}\delta_{is}\}$.

Given a simple undirected graph G, then by substituting the above tensors into (7) and taking account on symmetries we conclude that any arbitrary linear permutation invariant function $Z(\mathbf{A})$ must be of the following form

$$Z(\mathbf{A})_{ij} \quad a_1 + \delta_{ij}\left(a_2 + a_3 k_j\right) + a_4 A_{ij}, \tag{10}$$

with $k_j \quad \deg_G(j)$ and $a_{1,2,3,4}$ being arbitrary constants.

If we require that the linear function Z preserves the notion of connectivity,

$$k_i \quad \sum_{j\in V} Z(\mathbf{A})_{ij}, \tag{11}$$

it is clear that we should take $a_1 \quad a_2 \quad 0$ (indeed, the contributions $a_1 N$ and a_2 are incompatible with (11)) and then obtain a relation for the remaining constants, $1 - a_3 \quad a_4$. Introducing the new parameter $\beta \equiv a_4 > 0$, we reformulate (10) as follows,

$$Z(\mathbf{A})_{ij} \quad (1-\beta)\,\delta_{ij} k_j + \beta A_{ij}. \tag{12}$$

If we express (11) in the form of a probability conservation relation, then the function $Z(\mathbf{A})$ acquires a probabilistic interpretation. Substituting (12) back into (11), we obtain

$$\begin{aligned} 1 \quad & k_i^{-1} \sum_{j\in V} Z(\mathbf{A})_{ij} \\ & \sum_{j\in V} (1-\beta)\,\delta_{ij} + \beta \tfrac{A_{ij}}{k_i} \\ & \sum_{j\in V} T_{ij}^{(\beta)}. \end{aligned} \tag{13}$$

The operator $T_{ij}^{(\beta)}$ is nothing else as the generalized random walk transition operator if $0 < \beta \le k_{\max}^{-1}$ where $k_{\max}$ is the maximal degree in the graph G. In the "lazy" random walks defined by $T_{ij}^{(\beta)}$, a random walker stays in the initial vertex with probability $1-\beta$, while it moves to another node randomly chosen among its nearest neighbors with probability β/k_i. If we take $\beta \quad 1$, then the operator $T_{ij}^{(\beta)}$ describes the usual random walks extensively investigated in the classical surveys [14]-[16].

Being defined on a connected undirected graph, the matrix $T_{ij}^{(\beta)}$ is a real positive stochastic matrix, and therefore, in accordance to the Perron-Frobenius theorem [17], its maximal eigenvalue equals 1, and it is simple. A left eigenvector,

$$\pi T^{(\beta)} \quad \pi, \tag{14}$$

associated with the maximal eigenvalue 1 has positive entries satisfying the normalization condition (2) independently of the value β,

$$\pi_i \quad \frac{k_i}{\sum_{i\in V} k_i}. \tag{15}$$

It is interpreted as the unique equilibrium state π (the stationary distribution of random walks). If the graph G is not bipartite, any density σ ($\sigma_i > 0$, $\sum_{i\in V} \sigma_i = 1$) asymptotically tends to the stationary distribution under the action of the transition operator $T_{ij}^{(\beta)}$,

$$\pi = \lim_{t\to\infty} \sigma \left(T^{(\beta)}\right)^t. \tag{16}$$

If G is regular, then π is uniform. For a non-regular graph G, this property is replaced by time reversibility (the balance equation)

$$\pi_i T_{ij}^{(\beta)} = \pi_j T_{ji}^{(\beta)}, \quad i, j \in V. \tag{17}$$

The stationary distribution of random walks π is the unique density associated to the automorphisms of a simple undirected graph G.

SMALL WORLDS OF BIG CITIES

The main focus of the space syntax study is on the relative proximity (or accessibility) between different locations and associating these distances with densities and intensities of human activity which occur at different open spaces and along the links which connect them [18, 19, 20]. The decomposition of the city space into a complete set of intersecting open spaces characterized by the certain traffic capacities produces a spatial networks which we call the dual graph representation of a city.

While identifying the spaces of motion which play the role of nodes in the dual graphs of the compact city patterns bounded by natural geographical limitations, in the present paper we implement the street named oriented identification principle, the same which we used in our previous study [21]. The street named approach has been previously used by different authors, [23]-[26]. We have generalized their approach in a way to account the possible discontinuities of streets. Namely, we assign an individual street identification code (ID) to each continuous part of a street even if all of them share the same street name. Then the dual graph of the urban pattern is constructed by mapping spaces of motion coded with the same ID into nodes of the dual graph and intersections among each pair of individual spaces into edges connecting the corresponding nodes of the dual graph.

The graph theoretic distance (also: depth) d_{ij} between two locations in the urban pattern is the least number of steps (or the elementary navigation actions) required to reach the node i from the node j in the dual graph representation of the city. In particular, if two continuous streets intersect, then the distance between them equals one. Given the dual graph representation of a city, we can compute the distances between all pairs of its nodes.

The number of neighbors a node has at a distance d in the dual graph quantifies the relative structural importance of the location in the urban environment, [3]. The statistics of far away neighbors can be presented in the form of cumulative distribution functions $\mathfrak{P}_d(n)$ quantifying the probabilities that a node has at least n neighbors at a distance d. Being monotonous functions of n, cumulative distributions reduce the noise in the distribution tails, however the adjacent points on their plots are not statistically independent.

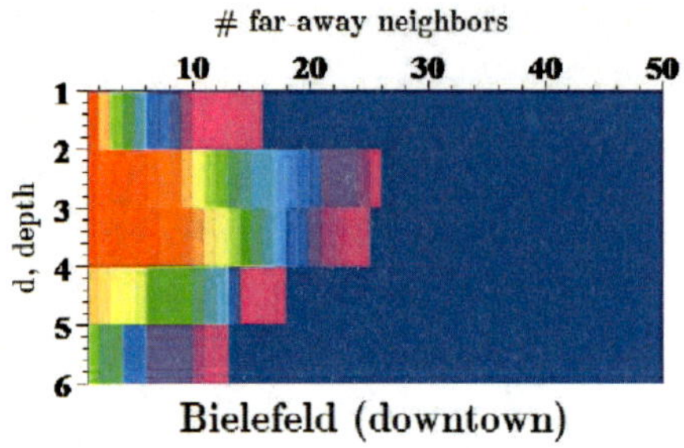

FIGURE 1. The cumulative distributions of far-away neighbors in the dual graph of Bielefeld downtown. The cumulative degree distribution are shown in the first row. The cumulative distributions of far-away neighbors are encoded in the second and forthcoming rows. Probability is ranked from zero (dark blue) to 1 (red).

For regular graphs, the number of far-away neighbors grows up exponentially fast with the distance. In Figs. 1, 2, we have displayed the cumulative distribution functions $P_d(n)$ quantifying the probabilities of that an average street is interconnected with at least n other streets within a distance d calculated for the dual graphs of two German organic cities - the downtown of Bielefeld in Westphalia and Rothenburg ob der Tauber in Bavaria. It is obvious that most of the streets in these towns can be reached in three navigation steps. The compactness of German medieval towns (see Figs. 1, 2) uncovers the historical functional background beyond their structure.

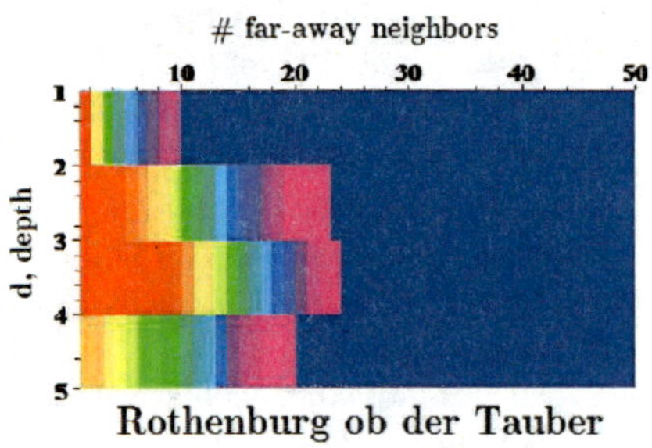

FIGURE 2. The cumulative distributions of far-away neighbors in the dual graphs of Rothenburg ob der Tauber, Bavaria. The cumulative degree distribution are shown in the first row. The cumulative distributions of far-away neighbors are encoded in the second and forthcoming rows. Probability is ranked from zero (dark blue) to 1 (red).

The term of a *burgh* has been in use since the 12th century, when David I of Scotland created the first Royal burghs. Early burghs were granted the power to trade, which allowed them to control trade until the 19th century, [27]. Their spatial structure was shaped by the public activities such as trade and exchange - ordered in such a way as to maximize the presence of people in central areas. In such an *organic* city, the majority of streets are just by a few syntactic steps away from each other, so that the entire city is compact that can be clearly seen from statistics of far away neighbors shown in Figs. 1-4.

It is interesting to mention that in the larger urban patterns such as the street grid in Manhattan and the canal network of Venice, most of the nodes of spatial networks can still be reached in less as three navigation steps (see. Figs. 3,4). However, there are also noticeable fractions of relatively isolated places. The structure of cumulative distributions shown in Figs. 3, 4 suggests that the correspondent graphs contain many cliques (complete subgraphs), together with subgraphs being "almost" cliques. In complex network theory, this phenomenon is called the *small world* property. Small-world networks are characterized by a small diameter and a high clustering coefficient having connections between almost any two nodes within them. Hubs - nodes in the network serving as the common connections mediating the short path lengths between other edges are commonly associated with small-world networks, [28].

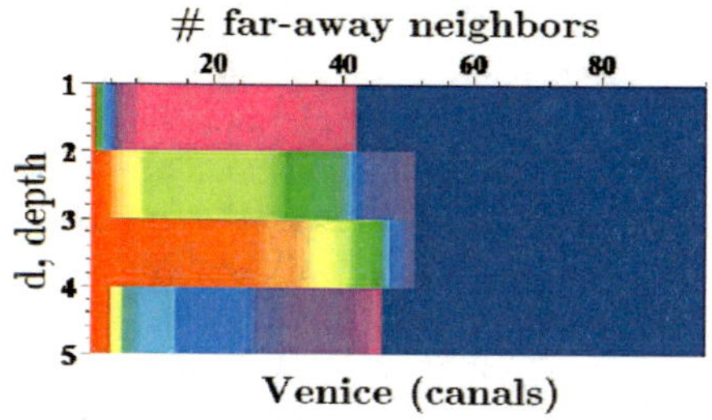

FIGURE 3. The cumulative distributions of far-away neighbors in the dual graph of Venetian canal network (96 canals). Probability is ranked from zero (dark blue) to 1 (red).

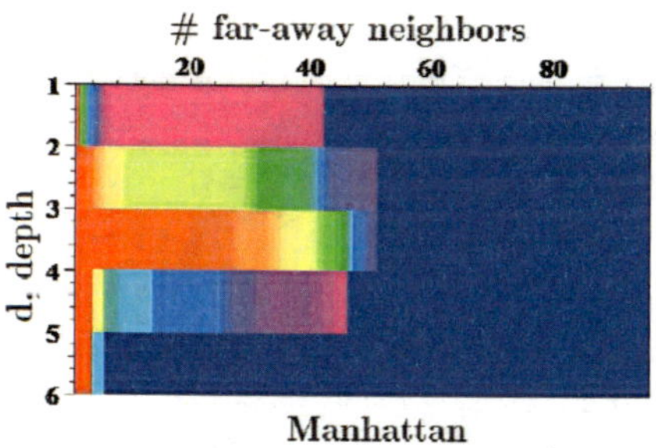

FIGURE 4. The cumulative distributions of far-away neighbors in the dual graph of the street grid in Manhattan. Probability is ranked from zero (dark blue) to 1 (red).

The tendency to shorten distances in urban space networks induced by the public processes is complemented in the "small world" cities (like those represented in Fig. 3, 4) with the residential process which shapes relations between inhabitants and strangers preserving the original residential culture against unsanctioned invasion of privacy. While the majority of streets and canals characterized with an excellent accessibility promotes commercial activities and intensifies cultural exchanges, certain districts of such cities demonstrate the opposite tendency having residential areas relatively segregated from the rest of the city, [3].

The main technical problem of space syntax investigation is to rank out the open spaces in the city in accordance to their accessibility.

FROM RANDOM WALKS TO EUCLIDEAN SPACES

The idea of using the spectral properties of self-adjoint operators in order to extract information about graphs is standard in spectral graph theory [29] and in theory of random walks on graphs [14]-[16]. In the following calculations, in the transition operator (13) we take $\beta = 1$ that allows to compare directly the forthcoming formulas with those known from the classical surveys on random walks defined on graphs, [14]-[16].

Among all measures which can be defined on the set of nodes V, there is one associated to the stationary distribution of random walks, $m = \sum_{i\in V} \pi_i \delta_i$, in which δ_i is the vector of the canonical basis that equals 1 at $i \in V$ and zero otherwise, with respect to which the transition operator ((13) with $\beta = 1$) is self-adjoint in the Hilbert space $\mathscr{H}(V)$,

$$\widehat{T} = \frac{1}{2}\left(\pi^{1/2} T \pi^{-1/2} + \pi^{-1/2} T^{\top} \pi^{1/2}\right). \tag{18}$$

In the above equation, $T^{\top}$ is the adjoint operator, and π is considered as the diagonal matrix diag $(\pi_1, \ldots, \pi_N)$. In particular, the elements of the symmetric transition operator defined on simple undirected graphs equal

$$\widehat{T}_{ij} = \frac{1}{\sqrt{k_i k_j}}, \quad \text{iff } i \sim j, \tag{19}$$

and zero otherwise.

The orthonormal ordered set of real eigenvectors ψ_i, $i = 1 \ldots N$, of the symmetric transition operator $\widehat{T}$ forms a basis in Hilbert space $\mathscr{H}(V)$. The components of the first eigenvector ψ_1 belonging to the largest eigenvalue $\mu_1 = 1$,

$$\psi_1 \widehat{T} = \psi_1, \quad \psi_{1,i}^2 = \pi_i, \tag{20}$$

describes the connectivity of nodes ($\pi_i \propto k_i$). The Euclidean norm of the orthogonal complement of ψ_1, $\sum_{s=2}^{N} \psi_{s,i}^2 = 1 - \pi_i$, quantifies the probability that a random walker is not in i. The eigenvectors, $\{\psi_s\}_{s=2}^{N}$, belonging to the eigenvalues $1 > \mu_2 \geq \ldots \mu_N \geq -1$ describe the connectedness of the graph G. The exterior products of vectors calculated by means of determinants are used in Euclidean geometry to study areas, volumes, and their higher-dimensional analogs, [30]. Denote by C_m the set of all subsets of V containing precisely $1 \leq m \leq N$ nodes. It is obvious that the size $|C_m|$ is given by the binomial coefficient,

$$|C_m| = \frac{N!}{m!\,(N-m)!}.$$

Let us note that the eigenvectors ψ_i, $i = 1 \ldots N$, induce basis systems for the spaces of all square summable m-point functions $f : \{i_1, \ldots i_m \in V\} \to \mathbb{R}$. We consider the matrix of eigenvectors of the symmetric transition operator (20),

$$\Psi \quad \begin{pmatrix} \psi_{1,1} & \cdots & \psi_{1,N} \\ \vdots & \vdots & \vdots \\ \psi_{N,1} & \cdots & \psi_{N,N} \end{pmatrix}, \tag{21}$$

in which the rows $\{\psi_k\}_{k=1}^{N}$ are orthonormal, $(\psi_i, \psi_j) \quad \delta_{ij}$. It is well known that $\det\Psi \quad \pm 1$ and numerically equal to the content of the parallelotope spanned by the unit eigenvectors $\{\psi_k\}_{k=1}^{N}$.

The nice property of minors $\Psi_{s_1,\ldots s_m}^{i_1,\ldots i_m}$ of order m obtained by the deletion of m rows $\{s_l\}$, $l \quad 1,\ldots m$, and columns $\{i_l\}$, $l \quad 1,\ldots m$, from $\det\Psi$ is that they are equal to their algebraic complements in Ψ, multiplied by ± 1, [31]. Moreover, the complete set of minors of order m forms an orthonormal system of $m-$point antisymmetric functions,

$$\sum_{i_1,\ldots i_m} \Psi_{s_1,\ldots s_m}^{i_1,\ldots i_m} \Psi_{s_1',\ldots s_m'}^{i_1,\ldots i_m} \quad \delta_{s_1 s_1'} \ldots \delta_{s_m s_m'}, \tag{22}$$

and can therefore be chosen as the basis in the Euclidean space $\mathbb{R}^{|C_m|}$, [31]. In the next section, we use the minors of order one to construct the embedding of undirected graphs into Euclidean space. In particular, we demonstrate that since any density σ taken as an initial distribution of random walkers on an undirected graph converges to the unique stationary distribution π as $t \to \infty$, this Euclidean space is $\mathbb{R}^{N-1}$.

The orthonormal systems of functions (22) are useful for decomposing the data when analyzing the traffic properties of transport networks. The data (say, of the real-time traffic flows and speed measurements) from m different detectors taken at different time slices can be uniquely decomposed using the orthonormal system (22) into smaller, more manageable parts and then compared allowing therefore an effective classification of traffic speed patterns.

EUCLIDEAN EMBEDDING OF GRAPHS BY RANDOM WALKS

In the present section, we consider the orthogonal system of minors of order one, Ψ_s^i, which are numerically equal to the correspondent components of the eigenvectors $\psi_{s,i}$. In particular, it is clear that the minors $\left(\Psi_1^i\right)_i^2 \quad \pi_i$ in accordance to (20). The minors of order one ($m \quad 1$) are associated to the nodes of the graph and provide the basis for the Hilbert space $\mathscr{H}(V)$. We demonstrate that a simple undirected graph can be embedded into the $(N-1)$ Euclidean space, so that for any pair of nodes the distance and angle can be uniquely defined.

It is important to note that Markov's symmetric transition operator $\widehat{T}$ defines a projection of any density $\sigma \in \mathscr{H}(V)$ on the eigenvector ψ_1 related to the stationary distribution π,

$$\sigma\widehat{T} \quad \psi_1 + \sigma_\perp \widehat{T}, \quad \sigma_\perp \quad \sigma - \psi_1, \tag{23}$$

in which $\sigma_\perp$ is the vector belonging to the orthogonal complement of ψ_1. It follows from (16) that $\lim_{t\to\infty} \sigma_\perp \widehat{T}^t \quad 0$, so that the vector $\sigma_\perp$ characterizes the transient process induced by σ.

In space syntax, we are interested in a comparison between the different densities defined on the nodes of the graph. It is convenient to compare them in accordance to their proximity to the unique stationary distribution of random walks π defined on the graph G. Since all components $\psi_{1,i} > 0$, we can rescale the density σ by dividing its components by the components of ψ_1,

$$\widetilde{\sigma}_i \quad \frac{\sigma_i}{\psi_{1,i}} \quad \frac{\sigma_i}{\sqrt{\pi_i}}. \tag{24}$$

It is clear that any two rescaled densities $\widetilde{\sigma}, \widetilde{\rho} \in \mathscr{H}$ differ with respect to random walks only by their dynamical components orthogonal to ψ_1,

$$(\widetilde{\sigma} - \widetilde{\rho})\, \widehat{T}^t \quad (\widetilde{\sigma}_\perp - \widetilde{\rho}_\perp)\, \widehat{T}^t,$$

for all $t > 0$.

Therefore, we can define the distance $\| \ldots \|_T$ based on random walks between any two densities by

$$\| \sigma - \rho \|_T^2 \quad \sum_{t \geq 0} \left\langle \widetilde{\sigma}_\perp - \widetilde{\rho}_\perp \middle| \widehat{T}^t \middle| \widetilde{\sigma}_\perp - \widetilde{\rho}_\perp \right\rangle. \tag{25}$$

or, using the spectral representation of $\widehat{T}$,

$$\begin{aligned} &\| \sigma - \rho \|_T^2 \\ &\quad \sum_{t \geq 0} \sum_{s=2}^{N} \mu_s^t \langle \widetilde{\sigma}^\perp - \widetilde{\rho}^\perp | \psi_s \rangle \langle \psi_s | \widetilde{\sigma}^\perp - \widetilde{\rho}^\perp \rangle \\ &\quad \sum_{s=2}^{N} \frac{\langle \widetilde{\sigma}_\perp - \widetilde{\rho}_\perp | \psi_s \rangle \langle \psi_s | \widetilde{\sigma}_\perp - \widetilde{\rho}_\perp \rangle}{1 - \mu_s}, \end{aligned} \tag{26}$$

where we have used Dirac's bra-ket notations especially convenient for working with inner products and rank-one operators in Hilbert space.

If we introduce a new inner product for densities $\sigma, \rho \in \mathscr{H}(V)$ by

$$(\sigma, \rho)_T \quad \sum_{s=2}^{N} \frac{\langle \widetilde{\sigma}_\perp | \psi_s \rangle \langle \psi_s | \widetilde{\rho}_\perp \rangle}{1 - \mu_s}, \tag{27}$$

then (26) is nothing else but

$$\| \sigma - \rho \|_T^2 \quad \| \sigma \|_T^2 + \| \rho \|_T^2 - 2(\sigma, \rho)_T, \tag{28}$$

where

$$\| \sigma \|_T^2 \quad \sum_{s=2}^{N} \frac{\langle \widetilde{\sigma}_\perp | \psi_s \rangle \langle \psi_s | \widetilde{\sigma}_\perp \rangle}{1 - \mu_s} \tag{29}$$

is the square of the norm of $\sigma \in \mathscr{H}(V)$ with respect to random walks defined on the graph G.

We conclude the description of the $(N-1)$-dimensional Euclidean space structure of G induced by random walks by mentioning that given two densities $\sigma, \rho \in \mathscr{H}(V)$, the angle between them can be introduced in the standard way,

$$\cos \angle(\rho, \sigma) \quad \frac{(\sigma, \rho)_T}{\| \sigma \|_T \, \| \rho \|_T}. \tag{30}$$

Random walks embed connected undirected graphs into the Euclidean space $\mathbb{R}^{N-1}$. This embedding can be used in order to compare the accessibility properties of individual nodes and to construct the optimal coarse-graining representations of the transport networks.

Indeed, the vector δ_i of the canonical basis is the particular case of density, which is zero for all nodes excepting for the node $i \in V$ where it equals 1. In accordance to (29), the density δ_i acquires the norm $\| \delta_i \|_T$ associated to random walks defined on G. In the theory of random walks [14], its square,

$$\| \delta_i \|_T^2 \quad \frac{1}{\pi_i} \sum_{s=2}^{N} \frac{\psi_{s,i}^2}{1-\mu_s}, \tag{31}$$

gets a clear probabilistic interpretation expressing the spectral formula of access time to the node $i \in V$, the expected number of steps required for a random walker to reach the node i starting from an arbitrary node chosen randomly among all other nodes of the graph G with respect to the stationary distribution of random walks π. The squared norm of the canonical vector $\| \delta_i \|_T^2$ can be used in order to characterize the access to the node i from other nodes of the graph G. By the inequality between arithmetic and harmonic means, it is easy to prove [14] that

$$\| \delta_i \|_T^2 \geq \frac{(1-\pi_i)^2}{\pi_i}, \tag{32}$$

so that the nodes which are difficult to reach ($\pi_i \ll 1$) are also isolated from the rest of the graph.

The Euclidean distance between any two nodes of the graph G is calculated as the distance (26) induced by random walks between two canonical vectors δ_i and δ_j,

$$K_{i,j} \quad \| \delta_i - \delta_j \|_T^2 \quad \sum_{s=2}^{N} \frac{1}{1-\mu_s} \left(\frac{\psi_{s,i}}{\sqrt{\pi_i}} - \frac{\psi_{s,j}}{\sqrt{\pi_j}} \right)^2, \tag{33}$$

gives the spectral representation of commute time defined in the theory of random walks on undirected graphs as the expected number of steps required for a random walker starting at $i \in V$ to visit $j \in V$ and then to return back to i, [14].

The commute time which plays the role of the Euclidean distance between the nodes of the graph can be represented as a sum, $K_{i,j} \quad H_{i,j} + H_{j,i}$, in which

$$H_{i,j} \quad \| \delta_i \|_T^2 - (\delta_i, \delta_j)_T \tag{34}$$

is the first hitting time which plays probably the most important role in the quantitative theory of random walks [14]. It quantifies the expected number of steps before node j is visited if starting from node i. The first hitting time $H_{i,j}$ satisfies the equation

$$H_{i,j} \quad 1 + \frac{1}{k_i} \sum_{l \sim i} H_{l,j} \tag{35}$$

with the intimal condition $H_{i,i}$ 0. The equation (35) expresses the fact that the first step takes the random walker to a neighbor $l \sim i$ and then he has to reach j from there.

It is important to mention that the cosine of an angle calculated in accordance to (30) has the structure of Pearson's coefficient of linear correlations that reveals it's natural statistical interpretation. The notion of angle between any two nodes of the graph arises naturally as soon as we become interested in the strength and direction of a linear relationship between two random variables, the flows of random walks moving through them. If the cosine of an angle (30) is 1 (zero angles), there is an increasing linear relationship between the flows of random walks through both nodes. Otherwise, if it is close to -1, there is a decreasing linear relationship. The correlation is 0 if the variables are linearly independent. It is important to mention that as usual the correlation between nodes does not necessary imply a direct causal relationship (an immediate connection) between them.

PETERSEN GRAPH AND VENETIAN CANALS

In the present section, we construct and investigate the Euclidean embedding of two graphs. The first one we study is the Petersen graph with 10 nodes (see Fig. 5). Another example is the spatial network of 96 Venetian canals which serve the function of roads in the ancient city that stretches across 122 small islands (see Fig. 6). While identifying canals over the plurality of water routes on the city map of Venice, the canal-named approach has been used, in which two different arcs of the city canal network were assigned to the same identification number provided they have the same name. The

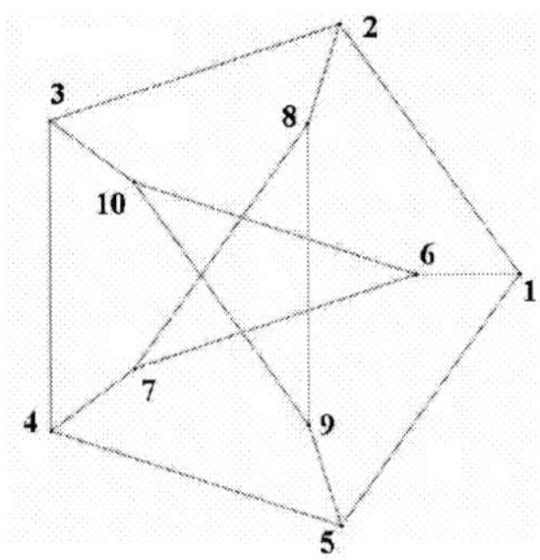

FIGURE 5. The Petersen graph.

Petersen graph is a regular graph, k_i $3, i$ $1, \ldots 10$, so that $\sum_i k_i$ 30, and the stationary distribution of random walks is uniform, $\pi_i^{(\mathrm{Pet})}$ 0.1. The spectrum of the random walk transition operator (18) defined on the Petersen graph contains the Perron eigenvalue μ_1 1 which is simple, then the eigenvalue μ_2 $1/3$ with multiplicity 5, and μ_3 $-2/3$ with multiplicity 4. Therefore, there are just 3 linearly independent eigenvectors, and two eigensubspaces for which the orthonormal basis vectors can be calculated, so that the resulting matrix of eigenvectors and basis vectors which we use in (31-33) always has full column dimension. Random walks embed the Petersen graph into a 9-dimensional Euclidean space, in which all nodes have equal norm (31), $\|i\|_T$ 3.14642654 meaning

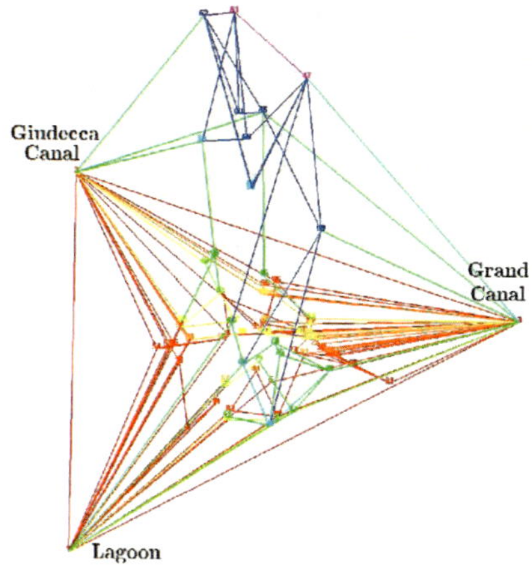

FIGURE 6. The dual graph representation of the spatial network of 96 Venetian canals

that the expected number of steps a random walker starting from a node chosen randomly with probability $p = 0.1$ reaches any node in the Petersen graph equals 9.9. Indeed,

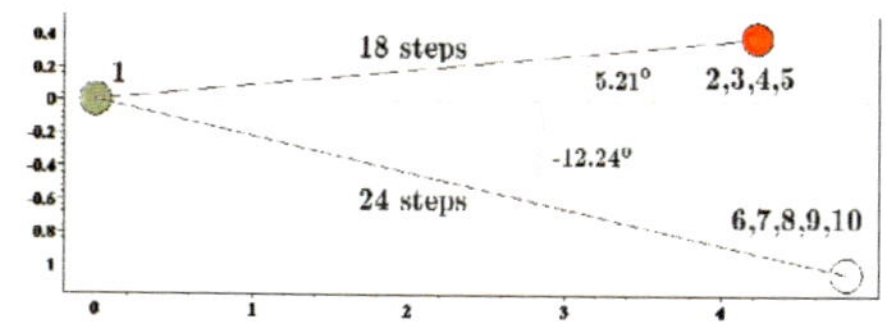

FIGURE 7. The Euclidean space embedding of the Petersen graph drawn with respect to the node #1.

the structure of 9-dimensional vector space induced by random walks defined on the Petersen graph cannot be represented visually, however if we choose one node as a point of reference, we can draw its 2-dimensional projection by arranging other nodes at the distances calculated accordingly to (33) and under the angles (30) they are with respect to the chosen reference node (see Fig. 7).

It is expected that a random walker starting at the node 1 visits any peripheral node $(2,3,4,5)$ and then returns back in 18 random steps, while 24 random steps are expected in order to visit any internal node in the deep of the graph $(6,7,8,9,10)$. It is also obvious that while the linear relationship between the random walks flows through the node 1 and those through any peripheral node is positive, it is negative with respect to the flows passing through the internal nodes. Due to the symmetry of the Petersen graph, the figure displayed on Fig. 7 is essentially the same if we draw it with respect to any other periphery node $(2,3,4,5)$. It is also important to note that it turns to be mirror-reflected if we draw it with respect to any internal node $(6,7,8,9,10)$. Therefore, we can conclude that the Petersen graph contains two groups of nodes, at the periphery and in the depth, which appears to be as much as a quarter more isolated from one another than the nodes within each group (18 random steps vs. 24 random steps). It is clear that the

9–dimensional embedding of the Petersen graph into Euclidean space is characterized by the highest degree of symmetry.

The dual graph of the spatial network of Venetian canals (Fig 6) is much more complicated than the Petersen graph. We construct it by mapping every canal into an individual node of the dual graph and connecting the pairs of nodes by arcs when these canals intersect. The resulting graph is far from being regular, so that the stationary distribution of random walks defined on it is not uniform. In [21], we have discussed that it is not evident if the degree distributions observed in compact urban patterns and in the Venetian canal network, in particular, follow a power law. The spectrum of the symmetric Markov transition operator (18) defined on that is presented in Fig. 8. The

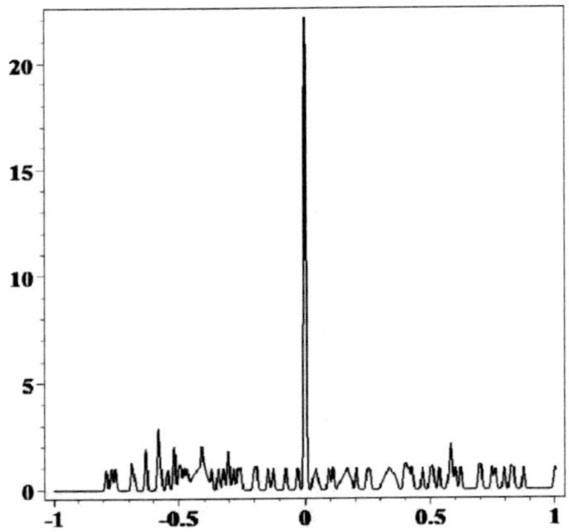

FIGURE 8. The spectrum of the Markov transition operator (19) defined on the spatial network of Venetian canals.

transition matrix $\widehat{T}$ for the canal network in Venice is strongly defective. In particular, it contains the eigenvalue $\mu = 0$ with multiplicity 22. This degeneracy indicates the presence of a complete bipartite subgraph in the spatial network of Venice shown in Fig. 6. In urban spatial networks encoded by their dual graphs, the values of access time (31) vary strongly from one canal to another being very long for topologically isolated places. Three data points characterized by the shortest access times shown in Fig. 9 are associated to the Lagoon of Venice, the Giudecca canal, and the Grand canal - the most central water routes in the city canal network. Four data points characterized by the worst accessibility represent the canal subnetwork of Venetian Ghetto.

The values of access times observed in the dual graph of Venetian canals scale with the connectivity of canals: the slope of the regression line equals 2.07.

A similar power law can be observed for the isolation patterns in other urban environments. For example, in Fig. 10, we have sketched the scatter plot of the connectivity vs. the norm a node in the dual graph representation of 50 streets in the downtown of Bielefeld. Professor Dr. Ludwig Streit works at the University of Bielefeld since 1972.

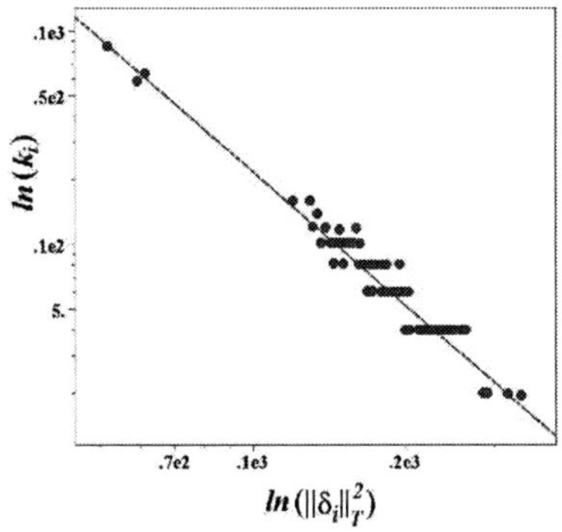

FIGURE 9. The scatter plot (in the log-log scale) of the connectivity vs. the value of access time to a node in the dual graph representation of 96 Venetian canals. Three data points characterized by the shortest access times represent the main water routes of Venice: the Lagoon of Venice, the Giudecca canal, and the Grand canal. Four data points of the worst accessibility are for the canal subnetwork of Venetian Ghetto. The slope of the regression line equals 2.07.

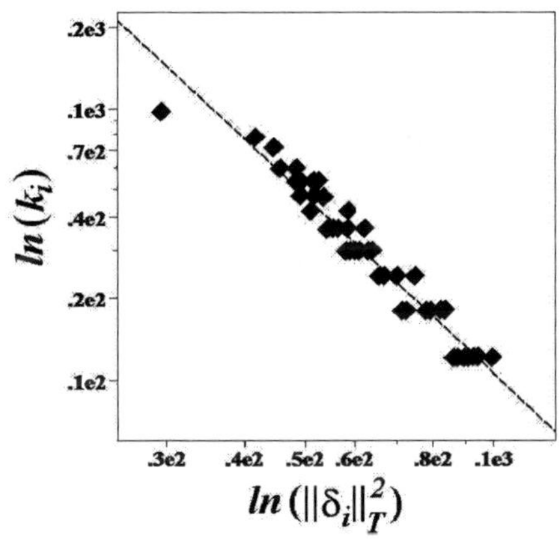

FIGURE 10. The scatter plot of the connectivity vs. the norm a node in the dual graph representation of 50 streets in the downtown of Bielefeld. The slope of the regression line equals 2.17.

In the space syntax analysis, the mean shortest distance from a given node i to any other node in the graph,

$$\ell_i \quad \frac{1}{N-1}\sum_{j\in V} d_{ij}, \tag{36}$$

is used for characterizing the level of integration of the node in the city pattern, [22]. The relation between the mean shortest distance to a node ℓ_i and the value of access time to the same node $\|\delta_i\|_T^2$ is very complicated and strongly depends on the topology of the graph. In Fig. 11, we have shown the scatter plot (in the log-log scale) of the

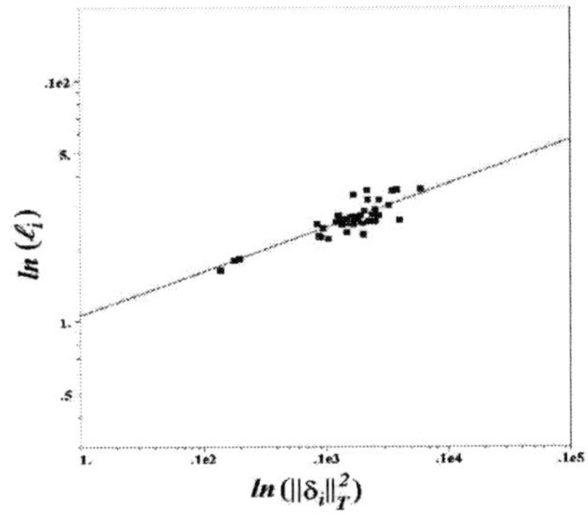

FIGURE 11. The scatter plot (in the log-log scale) of the mean shortest distance vs. the value of access time for the network of Venetian canals. The plot indicates a positive relation with the slope of regression line equals 0.18.

mean shortest distance vs. the value of access time for the network of Venetian canals. The plot Fig. 11 indicates a positive relation between ℓ_i and $\|\delta_i\|_T^2$ with the slope of the regression line equals 0.18.

The 2-dimensional projection of the Euclidean space of 96 Venetian canals set up by random walks drawn for the the Grand Canal of Venice (the point $(0,0)$) is shown in Fig. 12. Nodes of the dual graph representation of the canal network in Venice are shown by disks with radiuses taken equal to the degrees of the nodes. All distances between the chosen origin and other nodes of the graph (Fig. 6) have been calculated in accordance to (33) and (30) has been used in order to compute angles between nodes. Canals negatively correlated with the Grand Canal of Venice are set under negative angles (below the horizontal), and under positive angles (above the horizontal) if otherwise. It is evident

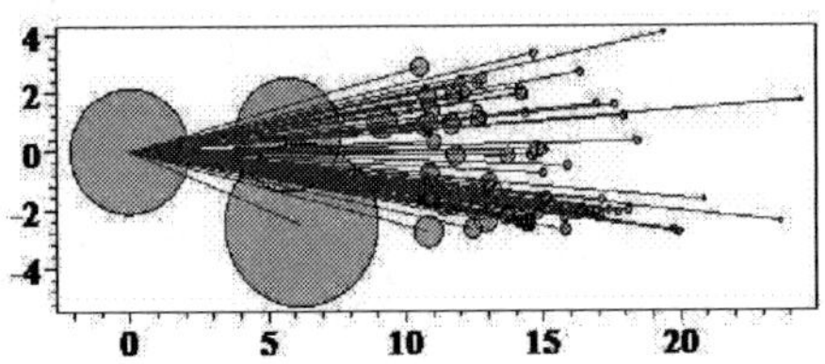

FIGURE 12. The 2-dimensional projection of the 95-dimensional Euclidean spaces associated to random walks defined on the city canal network built from the perspective of the Grand canal of Venice chosen as the origin. The labels of the horizontal axes display the expected number of random walk steps. The labels of the vertical axes show the degree of nodes (radiuses of the disks).

from Fig. 12 that disks of smaller radiuses demonstrate a clear tendency to be located far away from the origin being characterized by excessively long commute times with the reference node (the Grand canal of Venice), while the large disks which stand in Fig. 12 for the main water routes are settled in the closest proximity to the origin that intends an immediate access to them.

DISCUSSION AND CONCLUSION

From Euler's time, urban design and townscape studies were the sources of inspiration for the network analysis and graph theory. It is common now that networks are the reality of urban renewals [32]. Flows of pedestrians and vehicles through a city are dependent on one another and that requests for organizing them in a network setting. Moreover, the networking is structurally contagious. In order to be able to master a network effectively, an authority should also constitute a network structure, probably as complicated as the one that it supervises. A complex network of city itineraries that we can experience in everyday life appears as the result of multiple complex interactions between various transport, social, and economical networks.

In the present paper, we have developed a self-consistent approach for exploring the structure of urban environments based on the use of Markov chains (random walks) defined on the dual graph of the city. The stochastic transition matrix of random walks gives a matrix representation for the group of automorphisms of the graph $G(V,E)$. The expected number of steps required to reach a node of the graph starting from a node chosen from the stationary distribution over V given by (31) can be considered as a characteristic time scale $\mathscr{T}_{[i]}$ $\|\delta_i\|_T^2$ quantifying the accessibility of the node i in the graph G by random walks. The data of scatter plots displayed in Figs. 9,10 show that in the studied urban patterns $\mathscr{T}_{[i]}$ scales with the degree of vertex, so that $\mathscr{T}(k) \propto k^{-2-2\varepsilon}$, with some $\varepsilon > 0$ (ε 0.085 for the downtown of Bielefeld and ε 0.035 for the canal network of Venice). We have also discussed in Sec. that in the free flow regime of a queuing network preserving the proportion of time λ_i spent in $i \in V$ constant (which then can be considered as the normalized length distance of the given street i), the expected first passage time to the node from a node random chosen among all nodes of the queuing network equals λ_i^{-1}. While equalizing these passage times, we obtain that in the free flow regime the proportion of time λ spent by a moving agent in an open place of the given transport network scales with its connectivity as

$$\lambda(k) \propto k^{2+2\varepsilon}, \tag{37}$$

where the value $\varepsilon > 0$ is determined by the entire topology of the city.

It is well documented that isolation worsens an area's economic prospects by reducing opportunities for commerce, and engenders a feeling of isolation for the inhabitants, both of which can fuel poverty and crime [3]. The method we have introduced in the present paper could easily be used to identify isolated neighborhoods in big cities with a complex web of roads, walkways and public transport systems spotting hidden areas of geographical isolation in the urban landscape. The method can be used to analyze complex transport networks of any type.

ACKNOWLEDGMENTS

The support from the Volkswagen Foundation (Germany) in the framework of the project "*Network formation rules, random set graphs and generalized epidemic processes*" is gratefully acknowledged.

We wish to thank Thomas Küchelmann for his thorough reading of the paper and his excellent comments on that.

REFERENCES

1. Famous Quotations/Stories of Winston Churchill at http://www.winstonchurchill.org.
2. B. Hillier, J. Hanson, *The Social Logic of Space* (1993, reprint, paperback edition ed.). Cambridge: Cambridge University Press (1984).
3. B. Hillier, *The common language of space: a way of looking at the social, economic and environmental functioning of cities on a common basis*, Bartlett School of Graduate Studies, London (2004).
4. B. Hillier, S.Iida, "Network and psychological effects in urban movement", in A.G. Cohn, D.M. Mark (eds) *Proc. of Int. Conf. in Spatial Information Theory: COSIT 2005* published in *Lecture Notes in Computer Science* **3693**, 475-490, Springer-Verlag, (2005).
5. N. Biggs, E. Lloyd, and R. Wilson, *Graph Theory, 1736-1936*. Oxford University Press (1986).
6. M. Chown, "The Future Poverty Hiding In Cities", *The New Scientist*: 3 November 2007 (# 2628), London.
7. B. Hughes, *Random Walks and Random Environments*, Vol. 1, Sec. 2.1 (Oxford, 1995).
8. B. Tadic, "Exploring Complex Graphs by Random Walks", in P. Garrido and J. Marro (Eds.), *Modeling of complex systems: Seventh Granada Lectures*, Granada, Spain, AIP Conference Proceedings, **661**, pp. 24-26, Melville: American Institute of Physics, (2002).
9. Shi-Jie Yang, "Exploring complex networks by walking on them", *Phys. Rev. E* **71**, 016107, URL: http://link.aps.org/abstract/PRE/v71/e016107; doi:10.1103/PhysRevE.71.016107 (2005).
10. Luciano da Fontoura Costa and Gonzalo Travieso, "Exploring complex networks through random walks", *Phys. Rev. E* **75**, 016102; URL: http://link.aps.org/abstract/PRE/v75/e016102; doi:10.1103/PhysRevE.75.016102 (2007).
11. L. Breuer, D. Baum, *An Introduction to Queueing Theory*, Springer (2005).
12. S. Condamin, O Bénichou, V.Tejedor, R. Voituriez, J. Klaffer, First-passage times in complex scale-invariant media, *Nature* **450**, 77 -79; doi:10.1038/nature06201.
13. D.J. Aldous, "Lower Bounds for Covering Times for Reversible Markov Chains and Random Walks on Graphs", *J. Theor. Probab.* **2** (1) (1989).
14. L. Lovász, "Random Walks On Graphs: A Survey", *Bolyai Society Mathematical Studies* **2**: *Combinatorics, Paul Erdös is Eighty*, Keszthely (Hungary), p. 1-46 (1993).
15. L. Lovász, P. Winkler, *Mixing of Random Walks and Other Diffusions on a Graph.* Surveys in combinatorics, Stirling, pp. 119ï£¡154 (1995); London Math. Soc. Lecture Note Ser., vol. **218**, Cambridge Univ. Press.
16. D.J. Aldous, J.A. Fill, *Reversible Markov Chains and Random Walks on Graphs*. A book in preparation, available at www.stat.berkeley.edu/aldous/book.html.
17. R.A. Horn, C.R. Johnson, *Matrix Analysis*, Cambridge University Press, 1990 (chapter 8).
18. W. G. Hansen, *Journal of the American Institute of Planners* **25**, 73-76 (1959) .
19. A. G. Wilson, *Entropy in Urban and Regional Modelling*, Pion Press, London (1970).
20. M. Batty, *A New Theory of Space Syntax*, UCL Centre For Advanced Spatial Analysis Publications, CASA Working Paper
75 (2004).
21. Ph. Blanchard, D. Volchenkov, "Random Walks along the Streets and Channels in Compact Cities: Spectral analysis, Dynamical Modularity, Information, and Statistical Mechanics", *Phys. Rev. E* **75**, 026104 (2007). URL:http://link.aps.org/abstract/PRE/v75/e026104.
22. B. Jiang, "A space syntax approach to spatial cognition in urban environments", Position paper for NSF-funded research workshop *Cognitive Models of Dynamic Phenomena and Their Representations*,

October 29 - 31, 1998, University of Pittsburgh, Pittsburgh, PA (1998).
23. A. Cardillo, S. Scellato, V. Latora, and S. Porta, *Phys. Rev. E* **73**, 066107 (2006).
24. S. Porta, P. Crucitti, and V. Latora, *Physica A* **369**, 853 (2006)
25. S. Scellato, A. Gardillo, V. Latora, and S. Porta, *Eur. Phys. J. B* **50**, 221 (2006).
26. P. Crucitti, V. Latora, and S. Porta, *Chaos* **16**, 015113 (2006).
27. G.R. Stewart, *Names on the Land*. Boston: Houghton Mifflin Company (1967).
28. M. Buchanan, *Nexus: Small Worlds and the Groundbreaking Theory of Networks*. Norton, W. W. and Co., Inc. ISBN 0-393-32442-7 (2003).
29. F. Chung, *Lecture notes on spectral graph theory*, AMS Publications Providence (1997).
30. S. MacLane, G. Birkhoff, *Algebra*, AMS Chelsea (1999).
31. T. Muir, *Treatise on the Theory of Determinants*. Dover, New York, 1960 (revised and enlarged by W. H. Metzler).
32. M. Haffner, M. Elsinga, *Urban renewal performance in complex networks Case studies in Amsterdam North and Rotterdam South*, W16 ï£¡ Institutional and Organizational Change in Social Housing Organization in Europe, Int. Conference on Sustainable Urban Areas, Rotterdam (2007).

5th Jagna International Workshop: STOCHASTIC AND QUANTUM DYNAMICS OF BIOMOLECULAR SYSTEMS
3 - 5 January 2008, Research Center for Theoretical Physics, Central Visayan Institute Foundation, Jagna, Bohol, Philippines

WORKSHOP SCHEDULE

TIME	*January 3, Thursday*	*January 4, Friday* STREIT - Fest	*January 5, Saturday* WIEGEL - Fest	*January 6 Sunday*
8:30 – 9:15	REGISTRATION	**John R. Klauder,** "*Selecta from a life-long obsession with path integrals*"	**Toru Nakamura,** "*Time-development of the explosion of a stochastic process having a repulsive drift of polynomial growth*"	**8:00** OPTIONAL TOUR
9:15 – 10:00	OPENING CEREMONY	**M. Victoria Carpio-Bernido,** "*White noise path integrals in stochastic neurodynamics*"	**Hiroshi Ezawa,** "*Convergent WKB series – how can it be?*"	
10:00 – 10:30	*COFFEE BREAK*	*COFFEE BREAK*	*COFFEE BREAK*	
10:30 – 11:15	**Takeyuki Hida** "*Infinite dimensional space in stochastic analysis*"	**Philippe Blanchard,** "*Random graphs, Markov processes, and spectral graph theory: a way of looking at complex networks*"	**Christopher C. Bernido,** "*Modeling protein conformations: Wiegel's helical polymer and the Hida-Streit white noise path integral*"	
11:15 – 12:00	**Alfonso Albano,** "*Chaos, Boltzmann, Shannon and electroencephalography*"	**Martin Grothaus,** "*The Feynman integrand as a white noise distribution beyond perturbation theory*"	**Jose Perico Esguerra,** "*Biased and persistent random walks with geometrically shrinking steps*"	
12:00 – 13:30	LUNCH	LUNCH	LUNCH	
13:30 – 14:15	**Nelson Cue,** "*A physicist's engagement with biotechnology*"	**Maria João Oliveira,** "*Markov evolutions in the continuum*"	***Contributed Papers***	
14:15 – 15:00	**Gisela P. Concepcion,** "*Biological structures, interactions, function and behavior: research opportunities for physicists*"	**José Luís da Silva,** "*Kawasaki dynamics for continuous particle systems*"	**Frederik W. Wiegel,** "*Scaling features of violent conflicts*"	
15:00 – 15:30	*COFFEE BREAK*	GROUP PHOTO / *COFFEE BREAK*	*COFFEE BREAK*	
15:30 – 16:15	**Christopher Monterola,** "*Beneficial role of noise in neural networks functioning*"	***Contributed Papers***	**Roundtable Discussion**	
16:15 – 17:00	***Contributed Papers***	**Ludwig Streit,** "*Finding Communities in large bipartite networks*"	CLOSING	
17:00 – 18:00	***Tutorial Lecture*:** **Ludwig Streit,** "*Walking on a net: a glimpse at the spectral theory of graphs*"	***FREE / Tutorials***		
19:00	WELCOME DINNER			

PARTICIPANTS

Wilfredo V. Alangui
Department of Mathematics
College of Science
University of the Philippines, Baguio
Governor Pack Road
Baguio City 2600, Philippines

Geneliza C. Acas
Central Visayan Institute Foundation
Jagna, Bohol 6308, Philippines

Alfonso Albano
Physics Department
Bryn Mawr College, Pennsylvania, USA
e-mail: alfonso.albano@gmail.com

Jun Ando
Department of Applied Physics
School of Engineering
Osaka University, 2-1 Yamadaoka
Suita, Osaka 565-0871, Japan
e-mail: ando@ap.eng.osaka-u.ac.jp

Henry Aringa
Department of Physics
MSU-Marawi, Marawi City, Philippines
e-mail: henparinga@yahoo.com

Filchito Renee G. Bagsican
Department of Physics
MSU-Iligan Institute of Technology
Tibanga, Iligan City 9200, Philippines
e-mail: tjbagsican@gmail.com

Lyndon Bastatas
Department of Physics
MSU-Iligan Institute of Technology
Iligan City 9200, Philippines

Mary Grace Bato
National Institute of Physics
University of the Philippines, Diliman
Quezon City 1101, Philippines

Philippe Blanchard
Fakultät für Physik, Universtät Bielefeld
Bielefeld, Germany
e-mail: blanchard@physik.uni-bielefeld.de

Christopher C. Bernido
Research Center for Theoretical Physics
Central Visayan Institute Foundation
Jagna, Bohol 6308, Philippines
e-mail: cbernido@mozcom.com

Jinky B. Bornales
Department of Physics
MSU-Iligan Institute of Technology
Iligan City 9200, Philippines
e-mail: jbornales@yahoo.com

Kristine Joy E. Carpio
De La Salle University
Manila 1004, Philippines
e-mail: carpiok@dlsu.edu.ph

M. Victoria Carpio-Bernido
Research Center for Theoretical Physics
Central Visayan Institute Foundation
Jagna, Bohol 6308, Philippines
e-mail: cbernido@mozcom.com

Gisela P. Concepcion
Marine Science Institute
University of the Philippines, Diliman
Quezon City 1101, Philippines
e-mail: giselapc@compass.com.ph

Mark Nolan Confesor
Department of Physics
MSU-Iligan Institute of Technology
Iligan City 9200, Philippines

Nelson Cue
Physics Department
Hong Kong University of Science & Technology
Clear Water Bay, Kowloon, Hong Kong
e-mail: nelson.cue@gmail.com

Jonalizae P. Cuñado
Central Visayan Institute Foundation
Jagna, Bohol 6308, Philippines

Joy Lyn D. Dela Cruz
Department of Physics
MSU-Iligan Institute of Technology
Iligan City 9200, Philippines

Eduardo C. Cuansing, Jr.
Physics Department, De La Salle University
Manila 1004, Philippines
e-mail: cuansinge@dlsu.edu.ph

Guido P. David
Department of Mathematics
University of the Philippines, Diliman
Quezon City 1101, Philippines

Ma. Duvelaine Dumaluan
Panglao 6340, Bohol, Philippines

Jose Perico Esguerra
National Institute of Physics
University of the Philippines, Diliman
Quezon City 1101, Philippines

Mae A. Evardo
Central Visayan Institute Foundation
Jagna, Bohol 6308, Philippines

Hiroshi Ezawa
1-25-5 Honkomagome
Bunkyo-ku, Tokyo, Japan
e-mail: hiroshi.ezawa@q.email.ne.jp

Vicencia B. Galbizo
Central Visayan Institute Foundation
Jagna, Bohol 6308, Philippines

Pablita C. Galos
Central Visayan Institute Foundation
Jagna, Bohol 6308, Philippines

Rosemarie S. Gania
Department of Physics
MSU-Iligan Institute of Technology
Iligan City 9200, Philippines
e-mail: roch_87@yahoo.com

Kim A. Gargar
Department of Physics
Mapua Institute of Technology
Manila 1002, Philippines

Beverly Villarente-Gemao
Department of Physics
MSU-Iligan Institute of Technology
Iligan City 9200, Philippines

Enrico B. Gravador
Department of Physics
MSU-Iligan Institute of Technology
Iligan City 9200, Philippines
e-mail: ebgravador@gmail.com

Christine Joy Gravino
Department of Physics
MSU-Iligan Institute of Technology
Iligan City 9200, Philippines

Martin Grothaus
Mathematics Department
University of Kaiserslautern, P.O. Box 3049
67653 Kaiserslautern, Germany
e-mail: grothaus@mathematik.uni-kl.de

Takeyuki Hida
Hirabari Minami 2-903
Tenpaku-ku, Nagoya, Japan 468-0020
e-mail: thbty181@ybb.ne.jp

Elsie C. Jimenez
Department of Chemistry
College of Science
University of the Philippines, Baguio
Governor Pack Road, Baguio City 2600
Philippines

John R. Klauder
Department of Physics and Mathematics
University of Florida, Gainesville
FL 32611-8440 USA
e-mail: john.klauder@gmail.com

Neal Alfie Y. Lasta
Department of Physics
MSU-Iligan Institute of Technology
Iligan City 9200, Philippines
e-mail: nealalfielasta@yahoo.com

Normillah Lindao
Department of Physics
MSU-Iligan Institute of Technology
Tibanga, Iligan City 9200, Philippines

Regine A. Loberternos
Department of Materials Science & Engineering
University of the Philippines, Diliman
Quezon City 1101, Philippines
e-mail: bl.reagan@gmail.com

Lou Serafin M. Lozada
Institute of Mathematical Sciences & Physics
University of the Philippines, Los Baños
4031 Laguna, Philippines
e-mail: lsmlozada@gmail.com

Jansen V. Mendez
Tagbilaran City Science High School
M. Parras Extension, Tagbilaran City
Bohol, Philippines
e-mail: sir_jansen@yahoo.com

Priscilla S. Macansantos
Department of Mathematics
University of the Philippines, Baguio
Governor Pack Road
Baguio City 2600, Philippines

Jingle B. Magallanes
Department of Physics
MSU-Iligan Institute of Technology
Iligan City 9200, Philippines

Elizabeth P. Melchor
Far Eastern University
Nicanor Reyes Street
Sampaloc, Manila 1008, Philippines

Christopher Monterola
National Institute of Physics
University of the Philippines, Diliman
Quezon City 1101, Philippines
e-mail: cmonterola@gmail.com

Megumi Nakamura
Department of Chemistry
Yokohama National University
Yokohama, Japan

Toru Nakamura
Department of Mathematics
Sundai Preparatory School
Kanda-Surugadai, Chiyoda-ku
Tokyo 101-8318, Japan
e-mail: toru-nak@agate.plapla.or.jp

Alviu Rey Nasir
Department of Physics
MSU-Iligan Institute of Technology
Iligan City 9200, Philippines

Maria João Oliveira
CMAF, University of Lisbon
P-1649-003 Lisbon, Portugal
e-mail: oliveira@cii.fc.ul.pt

Crismar Patacsil
Department of Physics
University of the Philippines, Baguio
Governor Pack Road, Baguio City, Philippines
e-mail: cppatacsil@upb.edu.ph

Anthony Paul Ramirez
Department of Physics
MSU-Iligan Institute of Technology
Iligan City 9200, Philippines

Ruby Anna O. Rañoa
Central Visayan Institute Foundation
Jagna, Bohol 6308, Philippines

Rosario L. Reserva
Department of Physics
MSU-Iligan Institute of Technology
Iligan City 9200, Philippines

Ranzivelle Marianne L. Roxas
National Institute of Physics
University of the Philippines, Diliman
Quezon City 1101, Philippines

January T. Salas
Department of Hepatology and
Gene Therapy
Center for Applied Medical Research
University of Navarra, Pio XII SS
Pamplona 31008, Spain

Dante J. Sauquillo
Physics Department
Mapua Institute of Technology
Manila 1002, Philippines
e-mail: djsauquillo@mapua.edu.ph

José Luís da Silva
Centro de Ciencias Matemáticas
Universidade da Madeira
9000 Funchal, Madeira, Portugal
e-mail: luis@uma.pt

Ruelson Solidum
Department of Physics
Mindanao Polytechnic State College
CM Recto Lapasan
Cagayan de Oro City 9000, Philippines

Ludwig Streit
Centro de Ciencias Matemáticas
Universidade da Madeira
9000 Funchal, Madeira, Portugal
and Fakultät für Physik, Universtät Bielefeld
Bielefeld, Germany
e-mail: streit@pcserv.physik.uni-bielefeld.de

Jacqueline A. Teo
Asia Pacific College
No. 3 Humabon Place, Magallanes
Makati City 1232, Philippines
e-mail: jacqueline@apc.edu.ph

Ruben C. Umaly
Far Eastern University
Nicanor Reyes Sr. Street
Sampaloc, Manila 1008, Philippines
e-mail: rumaly@feu.edu.ph

Lara Gay U. Villaruz
Department of Physics
University of San Carlos
Talamban, Cebu City 6000, Philippines
e-mail: lvillaruz@yahoo.com

Roberto Vitancol
Physics Department
Mapua Institute of Technology
Manila 1002, Philippines

Frederik W. Wiegel
Institute of Theoretical Physics
University of Amsterdam
Valckenierstraat 65
1018 XE Amsterdam, The Netherlands
e-mail: f.w.wgl@hotmail.com

Author Index

SUBJECT INDEX